De Anza College

Haas CNC Mill & Lathe Programmer

Lynn J. Alton
Editor

"Haas CNC Mill & Lathe Programmer"
Lynn J. Alton (CreateSpace)
ISBN: 1453773797
EAN-13: 9781453773796

Copyright © 2009, Lynn J. Alton

All rights reserved.
No part of this publication may be reproduced,
stored in a retrieval system, or transmitted in any form;
electronic, mechanical, photocopying, scanning or recording method;
without prior written permission.

Introduction

In order to operate and program a CNC Mill or Lathe, a basic understanding of machining practices and a working knowledge of math is necessary.

"Haas CNC Mill & Lathe Programmer" designed to be used by both operators and programmers. It is intended to give the student a basic help in understanding CNC programs and there applications. It is not intended as an in-depth study of all ranges of machine use, but as a "Reference" for some common and potential situations facing the student CNC programmers and CNC operators. Much more training and information is necessary before attempting to program on the machine.

"Haas CNC Mill & Lathe Programmer"
is meant as a supplementary teaching aid for students at De Anza Collage taking the Manufacturing CNC classes. These classes are designed to teach students the basic skills of programming and operation of Haas CNC Mills & Lathes.

The information in this "Haas CNC Mill & Lathe Programmer" is reviewed regularly and any necessary changes will be incorporated in the next revision. This material is subject to change without notice. This training information is being supplied to all students that are learning to program and operate Haas CNC Mills & Lathes at De Anza College.

Users who call possible defects to the attention of the editor, or the omission of some matter that is considered to be of general value, often render a service to the entire MCNC classes at
De Anza College.

"Haas CNC Mill & Lathe Programmer"
The editor desirers to increase the usefulness of this book so all criticisms; suggestions about revisions, omissions, or inclusion of new material are welcome by the editor.

Table of Contents

Description	Page
Title Page	1
Introduction	2
Editor & Editors Notes	10
De Anza College	11
De Anza College, Manufacturing CNC	12
Haas Automation	14
De Anza College, Financial Aid & Scholarships	16
Haas CNC Mill:	17
CNC Mill, Coordinates	18
CNC Mill, Introduction	19
CNC Mill, Programming With Codes	20
CNC Mill, Program Format	21
CNC Mill, Definitions Within The Format	22
CNC Mill, Programs First Couple Of Lines	23
CNC Mill, Often Used Preparatory "G" & "M" Codes	24
CNC Mill, G-Codes	25
CNC Mill, Letter Address Codes	29
CNC Mill, M-Codes	33
CNC Mill, Program Structure	35
CNC Mill, Program Subroutine (Calling Subroutine With M98)	38
CNC Mill, Circular Interpolation, G02, G03	43
CNC Mill, Circular Pocket Milling, G12, G13	47
CNC Mill, Circular Planes, G17, G18, G19	52
CNC Mill, Return To Reference Point, G28	56
CNC Mill, Cutter Compensation, G40, G41, G42	57
CNC Mill, Tool Length Offset (TLO) G43, G44, G49	64
CNC Mill, Thread Milling	66
CNC Mill, Engraving, G47	70
CNC Mill, Scaling, G51	71
CNC Mill, Another Way To Return To Machine Zero, G53	73
CNC Mill, Rotation, G68	74
CNC Mill, Bolt Hole Commands, G70, 71, 72	77
CNC Mill, Repeat Command, L	80
CNC Mill, Canned Cycles (C/C)	81
CNC Mill, C/C H/S Peck Drilling, G73	82
CNC Mill, C/C L/H Tapping, G74	88
CNC Mill, C/C Bore, Stop, R-Out, G76	90
CNC Mill, C/C Bore, Stop, Shift, R-Out, G76, G77	92
CNC Mill, C/C Drill, G81	94
CNC Mill, C/C Spot Drill, Counterbore, G82	96

Table of Contents

Description	Page
Haas CNC Mills	
CNC Mill, C/C Normal Peck Drill, G83	98
CNC Mill, C/C R/H Tapping, G84	102
CNC Mill, C/C Bore In, Bore-Out, G85	104
CNC Mill, C/C Bore, Stop, R-Out, G86	106
CNC Mill, C/C Bore, M-Retract, G87	108
CNC Mill, C/C Bore, Dwell, M-Retract, G88	110
CNC Mill, C/C Bore, Dwell, Bore Out, G89	112
CNC Mill, Absolute & Incremental Position Command, G90, G91	114
CNC Mill, Enable Mirror Image, G101	115
CNC Mill, G/P Pocket Milling, G150	120
CNC Mill, Accuracy Control of Corners, G187	127
CNC Mill, Safety Procedures	128
Read Before Operating This Machine	129
Observe All Of The Warnings & Cautions	130
Unattended Operation	130
Machine Travels	131
CNC Mill, Quick Start Up Guide	133
Start Up And Zeroing-Out Machine	134
Jog Commands & Aligning a Vise	135
Finding & Storing Work Offsets	136
Loading Tools & Storing Tool Length Offsets	138
CNC Mill, 4-Axis & 5-Axis	141
Hass Rotary Table HRT 210 (4-Axis)	142
4-Axis Cylindrical Mapping G107	145
4-Axis Program With Subprograms	146
Creating Four & Five-Axis Programs	148
Simultaneous Rotation & Milling	148
Spiral Milling	149
Possible Timing Problems	151
5-Axis Programming Notes	151
Crash Recovery Procedure	153
5-Axis, G codes	154
5-Axis, C/C H/S Peck Drilling, G153	155
5-Axis, C/C Reverse Tapping, G155	157
5-Axis, C/C Drilling, G161	158
5-Axis, C/C Spot Drilling, G162	159
5-Axis, C/C Normal Peck Drilling, G163	160

Table of Contents

Description	Page
5-Axis, C/C Tapping, G164	161
5-Axis, C/C Boring, G165	162
5-Axis, C/C Bore, Stop, G166	163
5-Axis, C/C Bore, Dwell, G169	164
5-Axis, Non-Vertical Rigid Tapping, G174 CCW, G184 CW	165
De Anza College, Manual Machine Lab	166
De Anza College, CNC Lab Projects	168
EAPPRENTICE.NET	169
Haas CNC Lathes	170
CNC Lathe, Coordinates	171
CNC Lathe, Introduction	173
CNC Lathe, Programming With Codes	174
CNC Lathe, Program Format	175
CNC Lathe, Definitions Within The Format	176
CNC Lathe, Programs First Couple Of Lines	177
CNC Lathe, Safe Startup Line	178
CNC Lathe, Program Structure	180
CNC Lathe, Often Used Preparatory "G" & "M" Codes	182
CNC Lathe, G-Codes	183
CNC Lathe, Letter Address Codes	186
CNC Lathe, M-Codes	189
CNC Lathe, Program Structure	192
CNC Lathe, Rapid Position Command, G00	194
CNC Lathe, Linear Interpolation Command, G01	195
CNC Lathe, Circular Interpolation, G01, G02	197
CNC Lathe, Circular Interpolation, G01, G03	198
CNC Lathe, Circular Interpolation, G02, G03	199
CNC Lathe, Corner Rounding, G01	202
CNC Lathe, Corner Chamfering 45°, G01	203
CNC Lathe, Auto Chamfering, G01	204
CNC Lathe, Chamfering & Rounding, G01, G02, G03	205
CNC Lathe, TNR Compensation, Manually Calculating, G41, G42	207
CNC Lathe, TNR Compensation, Calculating Diagram, G41, G42	208
CNC Lathe, TNR Compensation, Manually Adding, G41, G42	209
CNC Lathe, TNR Compensation, 7 Steps, G41, G42	210
CNC Lathe, Tool Tip Orientation, G41, G42	211
CNC Lathe, Machine Cycles For Turning, Facing & Grooving	212

Table of Contents

Description	Page
Haas CNC Lathes	
CNC Lathe, OD / ID, Stock Removal, G71	213
CNC Lathe, Roughing, Finishing, G71, G70	214
CNC Lathe, Face Stock Removal, G72	215
CNC Lathe, Roughing, Finishing, G72, G70	216
CNC Lathe, Irregular Path Stock Removal Cycle, G73	218
CNC Lathe, High Speed Peck Drill Cycle, G74	219
CNC Lathe, End Face Grooving, G74	220
CNC Lathe, OD / ID Grooving, G75	223
CNC Lathe, Thread Cutting Cycle, Multiple Pass, G76	226
CNC Lathe, C/C Drill In, Rapid Out, G81	230
CNC Lathe, C/C Drill In, Dwell, Rapid Out, G82	231
CNC Lathe, C/C Deep Hole Peck Drill, G83	232
CNC Lathe, C/C Tapping, G84	234
CNC Lathe, C/C Reverse Tapping, G184	235
CNC Lathe, C/C Bore In, Bore Out, G85	236
CNC Lathe, C/C Bore In, Stop, Rapid Out, G86	237
CNC Lathe, C/C Bore In, Manual Retract, G87	238
CNC Lathe, C/C Bore In, Dwell, Manual Retract, G88	239
CNC Lathe, C/C Bore In, Dwell, Bore Out, G89	240
CNC Lathe, OD / ID Turning, G90	241
CNC Lathe, Thread Cutting Cycle, Model, G92	243
CNC Lathe, Thread Cutting, Chart	245
CNC Lathe, "D" Value Chart for O.D. Threads, 7 TPI through 14 TPI	246
CNC Lathe, "D" Value Chart for O.D. Threads, 16 TPI through 40 TPI	247
CNC Lathe, End Face Cutting Cycle, G94	248
Haas CNC Lathe Live Tooling	250
Live Tooling, M Codes	251
Live Tooling, Synchronous Milling	252
Live Tooling, Fine Spindle Control	253
Live Tooling, Installation	254
Live Tooling, VDI Adapter Installation	255
Live Tooling, C-Axis	256
Live Tooling, Cartesian to Polar	257
Live Tooling, Sample Programs, G05	258
Live Tooling, Sample Programs, G77	260
Live Tooling, Sample Programs, G112	263
Live Tooling, Sample Programs, G95, G186	264
Live Tooling, Sample Programs, G195, G196	264

Table of Contents

Description	Page
Haas CNC Lathe, Safety Procedures	265
Read Before Operating This Machine	266
Observe All Of The Warnings & Cautions	267
Unattended Operation	268
Lathe Specifications (SL-10 & SL-20)	269
Haas CNC Lathe, Quick Start Up Guide	271
Powering On The Machine	272
Creating A Program	272
Indexing A Tool In The Turret	272
Toughing Off Part To Get Tool Offsets	273
Toughing Off The Probe To Get Tool Offsets	274
Setting Work Zero Offsets Using A Probe	275
Running A Program	275
Haas CNC Lathe, Accessories	
CNC Lathe, Chuck Installation	276
CNC Lathe, Collet Installation	277
CNC Lathe, Drawtube Cover Plate	277
CNC Lathe, Re-Positioning Chuck Jaws	278
CNC Lathe, Auto Air Jet (optional)	278
CNC Lathe, Parts Catcher	279
CAD-RESOURCES.COM	281
Allowances & Fits	282
Allowances & Fits: Starrett	282
Allowances & Fits: ASA B4a-1925	286
Allowances & Fits: ANSA B4.1-1967	290
Drilling & Reaming	292
Drilled Hole Tolerances	292
Drilling Feed Table	292
Reaming Stock Allowance	292
Center Drill Dimensions, Std 60° Center Drill	293
Center Drill, Drill Point Allowance	293
Drill Point Depth & Countersink Allowance	295
Spotting Drills	295
Drill Point Angles	297
Decimal Equivalents, Drills	298

Table of Contents

Description	Page
Counterbore & Countersink	300
Countersinks, Blunt Ended	301
Counterbore, Socket Head Cap Screws, Inch	302
Countersink, Flat Head Screws, Inch	304
Counterbore, Socket Head Cap Screws, Metric	306
Countersink, Flat Head Screws, Metric	307
Tapping & Threading	308
Inch Cutting Tap: Tap Drill Size	309
Thread Gage Turns	309
Inch Roll-Form Tap (65%): Tap Drill Size	310
Metric Cutting Tap: Tap Drill Size	310
Metric Roll-Form Tap (65%): Tap Drill Size	310
Tread Measurement: Three-Wire Method	311
Tap & Drill Chart, Inch (Std & Roll Form)	312
Tap & Drill Chart, Metric (Std & Roll Form)	313
Metric Thread Pitch Conversion	314
Unified Coarse Helicoil Tap & Drill Chart	315
Unified Fine Helicoil Tap & Drill Chart	316
Metric Coarse, Helicoil Tap & Drill Chart	317
Metric Fine, Helicoil Tap & Drill Chart	318
Pipe Tap & Drill Chart	319
Pipe Sizes (Reference)	319
Feeds & Speeds	320
Feeds & Speeds, Formulas	321
Feeds & Speeds, Abbreviations	322
Chip Load Per Tooth for Mill Cutters	323
Drilling Speeds for Common Materials	323
Cast Iron – Annealed – Scale Removed	324
Brass & Bronze	324
Aluminum Alloys	324
Brinell Hardness Approximate:	325
Machinability Comparison	326
Surface Finishes (Roughness Average)	328
Strap Clamping Force	329
Tool Bits & Cutters	330
Relief Angles HSS Tool Bits	332
Chip Breakers	333
Corner Rounding Cutter	334

Table of Contents

Description	Page
Formulas, Northrop Grumman Calculator	335
Formulas, Conversions, English & Metric	336
Cutter Path Calculations	339
Tapers & Angles, Taper Per Foot & Taper Per Inch	343
Conversions, Angles, Degree – Minutes – Seconds	344
Conversions, Temperature, Celsius, Fahrenheit, Kelvin, Rankin	345
Properties of Bolt Circle	346
Properties of Circles	346
Properties of Circular Sector	347
Properties of Circular Segment	347
Properties of Circular Ring	348
Properties of Circular Ring Sector	348
Properties of Fillet (straight)	348
Trig, Right Triangle Relationships	349
Trig. Right Triangle	350
Trig. Oblique Triangle	351
Geometric Dimension Symbols	353
Drawings Symbols	358
Glossary	359
Haas CNC Notes	362
Exercises	366
Bolt Hole Location	367
Trig Cutter Path	370
Positioning, G90 Absolute & G91 Incremental	373
Program Interpolation	374
Program Plate, With Triangular Hole Pattern	375
Program Plate, With Three Grooves	377
Program Plate, Machined Both Sides, Side 1	379
Program Plate, With Bolt Circles	383
Program Plate, Machined Both Sides, Side 2	385
Program Happy Face	388
Editors, Suggested Reading	390
De Anza College, MCNC Program Course Description	391
De Anza College, MCNC Program Course Sequence	392

Lynn J. Alton – Editor

Dedication
The editor wishes to dedicate this "Hass CNC Mill & Lathe Programmer" to all his class-mates and instructors at De Anza College

Lynn has been self-employed since 1976 as an independent contractor and free-lance designer of PLC controlled machine tools, mechanical & electro-mechanical devices, Ultra-High Vacuum Systems and Components as well as High-Pressure Angioplasty Balloon Blowers. Lynn is also a successful project manager and was involved in high altitude research rockets in the late 1960's and has one patent to his credit and is co-inventor on another patent.

Lynn has the following degrees from De Anza College.

AA Degree: Liberal Arts – Science, Math & Engineering
AA Degree: Technical Writing.
AS Degree: Computer Aided Design.
AS Degree: Mfg. & CNC Product Model Making.
AS Degree: CNC Research & Development Machinist.

Editors Notes

Note: Material contained in this "Haas CNC Mill & Lathe Programmer" may be subject to copyrights belonging to the following companies.
Carr Lane Mfg. Co., St. Louis, MI
De Anza College, Cupertino, CA
Haas Automation, Inc., Oxnard, CA
Machinery's Handbook, NY
Northrop Grumman Corp.,

Note:
Wherever possible, permission has been obtained from the original copyright holders to use their material, but in many cases efforts to trace the owner of the copyrights have failed.

Note:
All images used in this "Book" have been used previously in the public domain by De Anza College or Haas Automation and are used here with their permission.

Note:
The editor would like to express his sincere appreciation to De Anza College and Haas Automation Inc. for allowing the use of their material in this "Haas CNC Mill & Lathe Programmer."

Note:
"Haas CNC Mill & Lathe Programmer" is intended for use in the following Manufacturing CNC classes at De Anza College.

MCNC 75A
CNC Programming, Beginning

MCNC 75B
CNC Programming, Intermediate

MCNC 75C
CNC Programming, Advanced

De Anza College

"Excellence" and "Innovation" are two words that have captured the essence of De Anza College for nearly four decades. Although each carries importance, together they're a winning combination that has established De Anza in the forefront of community colleges nationwide. Located 45 miles south of San Francisco, De Anza College occupies a 112-acre site in Cupertino in the heart of Silicon Valley. Cupertino is also home to Apple, Symantec, Sun, Hewlett-Packard, and many other high-technology firms. And is close to NASA at Moffett Filed, California

Nestled near the base of the Santa Cruz Mountains, the college was named after Spanish explorer Juan Bautista de Anza. The college is one of the largest, single-campus community colleges in the country with an enrollment averaging 25,000 students.

Operating on the quarter system, De Anza is one of two colleges in the Foothill-De Anza Community College District.

De Anza provides outstanding general education and vocational cour**ses** as well as interdisciplinary studies, community service opportunities, on-the-job-training, internships, collaborative programs with businesses and industries, and online and television classes. Find out what makes De Anza stand out from other colleges and what De Anza can offer you.

De Anza is committed to equal opportunity for all students and support students' freedom to express their beliefs. Please see our Anti-Discrimination Policy and principles of academic freedom for more information.

De Anza College
21250 Stevens Creek Boulevard
Cupertino, California 95014
www.deanza.edu

De Anza College
Manufacturing CNC

Would you like to work in a career that requires both mental and hands-on skills—doing creative high-precision work? Graduates in the Manufacturing & CNC fields are involved in forming and assembly of products from metals, plastics and composites. Sketches or computer-aided drafting (CAD) drawings are converted into three-dimensional parts, to very close tolerances, using a variety of manual and computer-aided machine tools (CNC). These products range from consumer items such as automobiles and home appliances to satellites and sophisticated medical equipment. Virtually everything purchased made of metal, plastic, or composites were ether produced directly, or with tools constructed by those in manufacturing and CNC careers.

The following represent some of the career specialties De Anza will qualify you for:

Research & Development Machinist - must have the same knowledge as the CNC Machinists. Additionally, they operate conventional (manually operated) lathes, milling machines and perform various abrasive machining operations such as grinding, honing and lapping. Has a working knowledge of a wide variety of manufacturing processes such as heat treating of materials and electro-discharge machining (EDM). Works closely with engineers and builds, assembles, and tests mechanical devices. Basically this person must be able to form almost any part from metals, plastics or composites.

CNC Machinist - operates and writes programs for computer numerical controlled (CNC) lathes and milling machines. Must be able to read blueprints or CAD files, select cutting tools and align the work-piece in the machine. Throughout the machining process the CNC machine is adjusted to deliver the required part dimensions and surface finishes while maintaining short cycle times. A good understanding of precision measurement is required for inspecting finished parts.

Other Advanced Careers:
Tool and Die Maker
Plastic Mold Maker
CNC Machine Programmer
Product Model Building
Manufacturing Systems Technician

De Anza College
Manufacturing CNC

Mike Appio; Department Head
Mark Larson: Full-Time Instructor
Andrew Stoddard: Technician
Larry Brown: Part-Time Instructor
Chris Newell: Part-Time Instructor
Mike Tatarakis: Part-Time Instructor
Derek Goodwin: Part-Time Instructor
Jim Mori: Part-Time Instructor

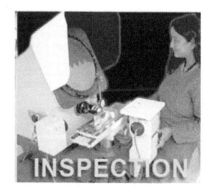

Haas Automation

Haas Automation, Inc.
Sturgis Road, Oxnard, California

The largest CNC machine tool builder in the Western World, Haas Automation manufactures a full line of CNC vertical and horizontal machining centers, CNC lathes, rotary tables and 5C indexers. Haas machine tools and rotary products are built to the exacting specifications of Gene Haas to deliver higher accuracy, repeatability and more durability than any other machine tools on the market.

Founded by Gene Haas in 1983, Haas Automation has always produced top-quality products at affordable prices.

By relying on volume sales rather than per-unit profits to build the company.

Haas Automation delivers more standard features, high-tech innovations and rock-solid engineering than perhaps any other CNC manufacturer in the world – and at better prices!

The machine builder firm continues to prosper, shipping a record 12,500 machines and gaining a sales volume of nearly $750-million, a 30% increase over 2005. Some 50% of machines are exported, and Bob Murray, General Manager of Haas Automation recently told a group of European distributors, that the company is targeting $1-billion in sales by 2010, with an export ratio of 70%.

Haas Automation

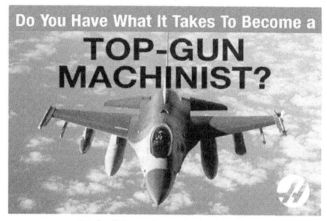

Gene Haas Foundation www.ghaasfoundation.org

De Anza Machinist Scholarship
this scholarship is intended for students in the Manufacturing / CNC Program

1) Applicant must have at least a 3.0 cumulative GPA from De Anza College.
2) Applicant must have completed at least 12 units in the MCNC Technology Program.
3) Applicant must be enrolled in at least 9 units at De Anza College.
4) Financial need will be considered.

To Apply: www.scholarships.fhda.edu

De Anza College
Financial Aid & Scholarships

NASA CAD Simulation Of Mars Rover

Financial Aid & Scholarships
www.deanza.edu/financialaid/scholarship
www.deanza.edu/financialaid

Veterans Scholarship Program
www.vets@scholarshipforveterans.org
www.lgl@cad-resources.com
www.foundation@fhda

NASA Paid Internships
www.internships.fhda.edu

Scholarship Giving
www.deanza.edu/financialaid/scholarshipgiving

Haas CNC Mill

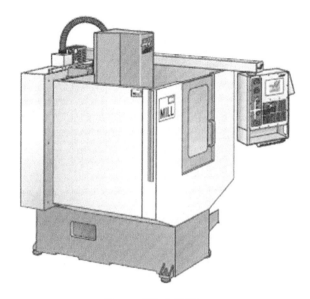

Haas Mini Mill

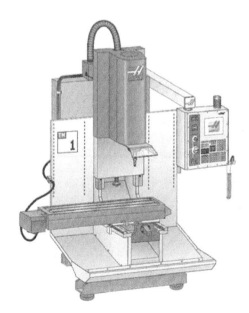

Haas Toolroom Mill

Haas CNC Mill Coordinates

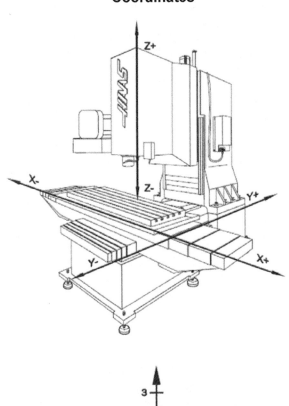

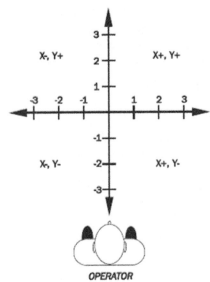

Haas CNC Mill
Introduction

This CNC manual provides basic programming principles necessary to begin programming the Haas CNC Milling Machine.

The CNC (Computerized Numerical Control) machine, the tool is controlled by a computer and is programmed with machine code system that enables it to be operated with minimal supervision and with a great deal of repeatability.

The same principles used in operating a manual machine are used in programming a CNC machine. The main difference is that instead of cranking handles to position a slide to a certain point, the dimension is stored in the memory of the machine control "once." The control will then move the machine to these positions each time the program is run.

In order to operate and program a CNC controlled machine, you must have a basic understanding of machining practices an a working knowledge of mathematics is necessary.

It is also important to become familiar with the control console and the placement of the keys, switches, displays, etc., that are pertinent to the operation of the machine.

This manual can be used for both operators and programmers. It is intended to give a basic understanding of CNC machine programming and it's applications. It is not intended as an in-depth study of all ranges of machine use, but as an overview of common and potential situations facing CNC programmers. Much more training and information is necessary before attempting to program on the machine.

This programming manual is meant as a supplementary teaching aid to users of the Haas Milling Machine. For a complete explanation and in-depth description, refer to the Programming and Operation Manual that is supplied with your Hass Milling Machine.

Manufacturing CNC, Rooms E21 – E26

Haas CNC Mill
Programming With Codes

The definition of a part program for any CNC consists of movements of the tool, and speed changes to the spindle RPM. It also contains auxiliary command functions such as tool changes, coolant on or off commands, or external M code commands.

Tool movements consist of rapid positioning commands, straight line moves or movement along an arc of the tool at a controlled speed.

The Haas mill has three (3) linear axes defined as X-axis, Y-axis and Z-axis.

The X and Y axis will move the machine table below and around the spindles centerline, while the Z axis moves the tool spindle down toward or up and

Away from the machine table. The machine-zero position is where the spindle is pointing down at the upper right corner, with the machine table all the way to the left in the X axis and all the way toward you in the Y axis and Z axis is up at the tool change position. Motion in the x axis will move the machine table to the right with negative values and to the left with positive values. The y axis will move the machine table toward you with positive values and away from you with negative values. Motion in the Z axis will move the tool toward the machine table with negative values and away from the machine table with positive values.

A program is written as a set of instructions given in the order they are to be performed. The instructions, if given in English, might look like this:

Line #1 = Select Cutting Tool

Line #2 = Turn Spindle On And Select The RPM

Line #3 = Rapid To The Starting Position Of The Part

Line #4 = Turn Coolant On

Line #5 = Choose Proper Feed Rate And Make The Cut (s)

Line #6 = Turn The Spindle And Coolant Off.

Line #7 = Return To Clearance Position To Select Another Tool and so on. But our machine control understands only these messages when given in machine code, also referred to as G and M codes programming. Before considering the meaning and the use of codes, it is helpful to lay down a few guidelines.

Haas CNC Mill
Program Format

There is no positional requirement for the address codes. They may be placed in any order within the block. Each individual can format their programs in many different ways. But, program format or program style is an important part of CNC machining. There are some program command formats that can be moved around, and some commands need to be a certain way, and there are some standard program rules that are just good to follow. The point is that a program needs to have an organized program format that's consistent and efficient so that any CNC machinist in your shop can understand it.

Some standard program rules to consider are:

Program X, Y, and Z in alphabetical order on any block. The machine will read Z, X or Y in any order, but we want to be consistent. If more than one of X, Y or Z is on a line, they should be listed together and in order. Write X first, Y next, then Z.

You can put G and M codes anywhere on a line of code. But, in the beginning when N/C programming was being developed G codes had to be in the beginning of a line and M codes had to be at the end. And this rule, a lot of people still follow and is a good standard to continue.

Some CNC machines allow you to write more than one M code per line of code and some won't. On the Haas, only one M code may be programmed per block and all M codes are activated or cause an action to occur after everything else on the line has been executed.

Program format is a series and sequence of commands that a machine may accept and execute. Program format is the order in which the machine code is listed in the program that consist of command words. Command words begin with a single letter and then numbers for each word. If it has a plus (+) value, no sign is needed. If it has a minus value, it must be entered with a minus (-) sign. If a command word is only a number and not a value, then no sign or decimal point is entered with that command. Program format defines the language of the machine tool.

N1 (mill outside edge) ;
T1 M06 (1/2 Dia. 4 flut end mill) ;
G90 G54 G00 X-2.3 Y2.3 S1600 M03 ;
G43 H01 Z0.1 M08 ;
G01 Z-0.625 F50. ;
G41 Y2. D01 F9.6 ;
X2. ;
Y-2. ;
X-2. ;
Y2.25 ;
G40 X-2.3 Y2.3 ;
G00 Z1. M09 ;
G28 G91 Y0. Z0. M05 ;
M01 ;

Haas CNC Mill
Definitions Within The Format

1) Character: A single alphanumeric character value or the "+" and "-" sign.

2) Word: A series of characters defining a single function such as a, "X" displacement, an "F" federate, or G and M codes. A letter is the first character of a word for each of the different commands. There may be a distance and direction defined for a word in a program. The distance and direction in a word is made up of a value, with a plus (+) or minus (-) sign. A plus (+) value is recognized if no sign is given in a word.

3) Block: series of words defining a single instruction. An instruction may consist of a single linear motion, a circular motion or canned cycle, plus additional information such as a federate or miscellaneous command (M-Codes).

4) Positive Signs: If the value following an address letter command such as A, B, C, I, J, K, R, U, V, W, X, Y, Z, is positive, the plus sign need not be programmed in. If it has a minus value it must be programmed in with a minus (-) sign.

5) Leading Zeros: if the digits preceding a number are zero, they need not be programmed in. The Haas control will automatically enter in the leading zeros.

Example: G0 for G00 and M1 for M01.
Trailing zeros must be programmed.
Example: M30 not M3, G70 not G7.

6) Model Commands: Codes that are active for more than the line in which they are issued are called "Model" commands. Rapid traverse, federate moves, and canned cycles are all examples of modal commands.

A "Non-Modal command which once called, are effective only in the calling block, and are then immediately forgotten by the control.

7) Preparatory Functions: "G" codes use the information contained on the line to make the machine tool do specific operations, such as:

a) Move the tool at rapid traverse.

b) Move the tool at a feed-rate along a straight line.

d) Move the tool along an arc at a feed-rate in a clockwise direction.

e) Move the tool along an arc at a feed-rate in a counterclockwise direction.

f) Move the tool through a series of repetitive operations controlled by "fixed cycles" such as, spot drilling, drilling, boring, and tapping.

8) Miscellaneous Functions: "M" codes are effective or cause an action to occur at the end of the block, and only one "M" code is allowed in each block of a program.

9) Sequence Numbers: N1 through N99999 in a program are only used to locate and identify a line or block and its relative position within a CNC program.

A program can be with or without Sequence Numbers. The only function of Sequence Numbers is to locate a certain block or line within a CNC program.

Haas CNC Mill
Programs First Couple Of Lines

The FIRST line or block in a program should be a tool number (T1) and a tool change (M06) command.

The SECOND line or block should contain an absolute (G90) command along with, a work-offset (G54 is the default), part zero command. A rapid (G00) command to position to an X, Y, coordinate location, a spindle speed command (Snnnn), and a spindle ON clockwise command (M03), or you could have the spindle speed and clockwise command defined on a separate line.

The NEXT line or block contains a "Read tool length compensation" command (G43), a tool length offset register number (H01), a Z-axis positioning move (Z1.0), and an optional coolant ON command (M08).

The tool startup lines with the necessary codes for each tool are listed below. These formats are a good example for the startup lines that are entered in for each tool.

T1 M06 (Text information)
G90 G54 G00 X0.5 Y-1.5 S2500 M03
G43 H01 Z1. M08

Another format you might choose is:

M06 T1 (Text information)
G00 G90 G54 X-1.5 Y2.5
S2500 M03
G43 Z1. H01 M08

Note:
A tool length offset number should usually always remain numerically matched with the tool number.

Setting 15 (the H & T code agreement) will ensure the tool number and the tool length offset number will match.

Example: T1 in line #1 should have H01 in line #3 or an alarm will occur if Setting 15 is on.

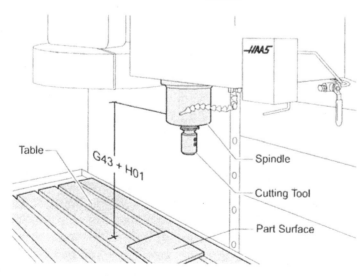

Haas CNC Mill
Often Used Preparatory "G" & "M" Codes

Often Used Preparatory "G" Codes

G00 – Rapid traverse motion; Used for non-cutting moves of the machine in positioning quick to a location to be machined, or rapid away after program cuts have been performed. Maximum rapid motion (IPM) of a Hass machine will vary on machine model.

G01 – Linear interpolation motion; Used fro actual machining and metal removal. Governed by a programmed federate in inches (or mm) per minute. Maximum feed-rate (IPM) of a Hass machine will vary on machine model.

G02 – Circular interpolation, Clockwise.

G03 – Circular interpolation, Counterclockwise.

G28 – Machine Home (Rapid Traverse)

G40 – Cutter Compensation CANCEL.

G41 – Cutter Compensation LEFT of the program path.

G42 – Cutter Compensation RIGHT of the program path.

G43 – Tool LENGTH Compensation +

G53 – Machine Coordinate Positional, Non-Model.

G54 – Work Coordinate #1
Part zero offset location.

G80 – Canned Cycle Cancel.

G81 – Drill Canned Cycle.

G82 – Spot Drill Canned Cycle.

G83 – Peck Drill Canned Cycle.

G84 – Tapping Drill Canned Cycle.

G90 – Absolute Programming Positioning.

G91 – Incremental Programming Positioning.

Often Used Preparatory "G" Codes

G98 – Initial Point Return Canned Cycle.

G99 – Rapid (R) Plane Return Canned Cycle.

Often Used Preparatory "M" Codes

M00 – Program Stop command on the machine. It stops the spindle, turns off coolant and stops look-a-head processing. Pressing CYCLE START again will continue the program on the next block of the program.

M01 – Optional Program Stop command. Pressing the OPT STOP key on the control panel signals the machine to perform a stop command when the control reads an M01 command. It will then perform like an M00.

M03 – Starts the spindle (CW) Clockwise. Must have a spindle speed defined.

M04 – Starts the spindle (CCW) Counterclockwise. Must have a spindle speed defined.

M05 – Stops the spindle

M06 – Tool change command along with a tool number will execute a tool change for that tool. This command will stop the spindle, Z-axis will move up to machine zero position and the selected tool will be put in the spindle. The coolant pump will turn off right before executing the tool change.

M08 – Coolant ON command.

M09 – Coolant OFF command.

M30 – Program End and Reset to the beginning of program.

M97 – Local Subroutine call.

M98 – Subroutine call.

M99 – Subprogram return.

Note: Only one "M" code per line.

Haas CNC Mill: G-Codes

1) G codes come in groups. Each group of G codes will have a specific group number.

2) A G code from the same group can be replaced by another G code in the same group. By doing this, the programmer establishes modes of operation. The universal rule here is that codes from the same group cannot be used more than once on the same line.

3) There are Modal G codes, which once established, remain effective until replaced with another G code from the same group.

4) There are Non-Modal G codes (Group 00) which once called, are effective only in the calling block, and are immediately forgotten by the control.

The first group (Group 01) controls the manner in which the machine moves. These moves can be programmed in either absolute or incremental. The codes are G00, G01, G02 and G03.

Each G code is a part of a group of G codes. The Group 0 codes are non-model; that is, they specify a function applicable to this block only and do not affect other blocks. The other groups are model and the specification of one code in the group cancels the previous code applicable from that group. A model G code applies to all subsequent blocks so those blocks do not need to re-specify the same G code.

There is also one case where the Group 01 G codes will cancel the group 9 (canned cycles) codes. If a canned cycle is active (G81 through G89), the use of G00 or G01 will cancel the canned cycle.

The rules above govern the use of the G codes used for programming the Haas Mills. The concept of grouping codes and the rules that apply will have to be remembered to effectively program the Haas Mill. The following is a list of Haas G codes. If there's a (Setting Number) listed next to a G code, that setting will in some way relate to that G code.

A single asterisk (*) indicates that it's the default G code in a group.
A double asterisk (**) indicates available options.

Haas CNC Mill: G-Codes

Code, Group, Function

G00, 01, * Rapid Positioning Motion (X, Y, Z, A, B) (Setting 10,56,101)
G01, 01, Linear Interpolation Motion (X, Y, Z, A, B, F)
G02, 01, Circular Interpolation Motion CW (X, Y, Z, A, I, J, K, R, F)
G03, 01, Circular Interpolation Motion CCW (X, Y, Z, A, I, J, K, R, F)
G04, 00, Dwell (P) (P=seconds "," milliseconds)
G09, 00, Exact Stop, Non-Model
G10, 00, Programmable Offset Setting (X, Y, Z, A, L, P, R)
G12, 00, Circular Pocket Milling CW (Z, I, K, Q, D, L, F)
G13, 00, Circular Pocket Milling CCW (Z, I, K, Q, D, L, F)
G17*, 02, Circular Motion XY Plane Selection (G02 or G02) (Setting 56)
G18, 02, Circular Motion ZX Plane Selection (G02 or G02)
G19, 02, Circular Motion YZ Plane Selection (G02 or G02)
G20*, 06, Verify Inch Coordinate Positioning
(Setting 9 will need to be INCH) (Setting 56)
G21, 06, Verify Metric Coordinate Positioning (Setting 9 will need to be METRIC)

Haas CNC Mill: G-Codes

Code, Group, Function

G28, 00, Machine Zero Return Thru Reference Point (X, Y, Z, A, B) (Setting 108)
G29, 00, Move to Location Thru G28 Reference Point (X, Y, Z, A, B)
G31, 00, ** Feed Until Skip Function (X, Y, Z, A, B, F)
G35, 00, ** Automatic Tool Diameter Measurement (D, H, Z, F)
G36, 00, ** Automatic Work Offset Measurement (X, Y, Z, A, B, I, J, K, F)
G37, 00, ** Automatic Tool Offset Measurement (D, H, Z, F)
G40, 07, * Cutter Compensation Cancel G41, G42, G141 (X, Y) (Setting 56)
G41, 07, 2D Cutter Compensation Left (X, Y, D) (Setting 43,44,58)
G42, 07, 2D Cutter Compensation Right (X, Y, D) (Setting 43,44,58)
G43, 08, Tool Length Compensation + (H, Z) (Setting 15)
G44, 08, Tool Length Compensation - (H, Z) (Setting 15)
G47, 00, Text Engraving (X, Y, Z, R, U, I, J, P, E, F)
(Macro Variable #599 to Change Serial Number)
G49, 08, * Tool Length Compensation Cancel G43, G44, G143 (Setting 56)
G50, 11, * Scaling G51 Cancel (Setting 56)
G51, 11, ** Scaling (X, Y, Z, P) (Setting 71)
G52, 12, Select Work Coordinate System G52 (Setting 33, YASNAC)
G52, 00, Global Work Coordinate System Shift (Setting 33, FANUC)
G52, 00, Global Work Coordinate System Shift (Setting 33, HAAS)
G53, 00, Machine Zero XYZ Positioning, Non-Modal
G54, 12, * Work Offset Positioning Coordinate #1 (Setting 56)
G55, 12, Work Offset Positioning Coordinate #2
G56, 12, Work Offset Positioning Coordinate #3
G57, 12, Work Offset Positioning Coordinate #4
G58, 12, Work Offset Positioning Coordinate #5
G59, 12, Work Offset Positioning Coordinate #6
G60, 00, Uni-Directional Positioning (X, Y, Z, A, B,) (Setting 35)
G61, 13, Exact Stop, Model (X, Y, Z, A, B)
G64, 13, * Exact Stop G61 Cancel (Setting 56)
G65, 00, ** Macro Sub-Routine Call
G68, 16, ** Rotation (G17, G18, G19) (X, Y, Z, A, R) (Setting 72,73)
G69, 16, * Rotation G68 Cancel (Setting 56)
G70, 00, Bolt Hole Circle with a Canned Cycle (I, J, L)
G71, 00, Bolt Hole Arc with a Canned Cycle (I, J, K, L)
G72, 00, Bolt Hole Along an Angle with a Canned Cycle (I, J, L)
G73, 09, High Speed Peck Drill Canned Cycle
(X,Y,A,B,Z,I,J,K, Q,P,R,L,F) (Setting 22)
G74, 09, Reverse Tapping Canned Cycle (X, Y, A, B, Z, R, J, L, F) (Setting 130,133)
G76, 09, Fine Boring Canned Cycle (X, Y, A, B, Z, I, J, P, Q, P, R, L, F) (Setting 27)
G77, 09, Back Bore Canned Cycle (X, Y, A, B, Z, I, J, Q, R, L, F) (Setting 27)
G80, 09, * Cancel Canned Cycle (Setting 56)

Haas CNC Mill: G-Codes

Code, Group, Function

G81, 09, Drill Canned Cycle (X, Y, A, B, Z, R, L, F)
G82, 09, Spot Drill, Counterbore Canned Cycle (X, Y, A, B, Z, P, R, L, F)
G83, 09, Peck Drill Deep Hole Canned Cycle (X, Y, A, B, Z, I, J, K, Q, P, R, L, F) (Setting 22,52)
G84, 09, Tapping Canned Cycle (X, Y, A, B, Z, R, J, L, F) (Setting 130,133)
G85, 09, Bore-in Bore-out Canned Cycle (X, Y, A, B, Z, R, L, F)
G86, 09, Bore-in Stop Bore-out Canned Cycle (X, Y, A, B, Z, R, L, F)
G87, 09, Bore-in Manual-Retract Canned Cycle (X, Y, A, B, Z, R, L, F)
G88, 09, Bore-in Dwell Manual-Retract Canned Cycle (X, Y, A, B, Z, P, R, L, F)
G89, 09, Bore-in Dwell Bore-Out Canned Cycle (X, Y, A, B, Z, R, L, F)
G90, 03, * Absolute Positioning Command (Setting 56)
G91, 03, Incremental Positioning Command (Setting 29)
G92, 00, Set Work Coordinate Value (Fanuc) (Haas)
G92, 00, Global Work Coordinate System Shift (Yasnac)
G93, 05, Inverse Time Feed Mode ON
G94, 05, * Inverse Time Feed Mode OFF, Feed Per Minute ON (Setting 56)
G95, 05, Feed Per Revolution
G98, 10, * Canned Cycle Initial Point Return (Setting 56)
G99, 10, Canned Cycle "R" Plane Return
G100, 00, Mirror Image Cancel
G101, 00, Mirror Image (X, Y, Z, A, B) (Setting 45,46,47,48,80)
G102, 00, Programming Output to RS-232 (X, Y, Z, A, B)
G103, 00, Limit Block Look-a-Head
(P0-P15 for number of lines control looks ahead)
G107, 00, Cylinder Mapping (X, Y, Z, A, Q, R)
G110, 12, Work Offset Positioning Coordinate #7
G111, 12, Work Offset Positioning Coordinate #8
G112, 12, Work Offset Positioning Coordinate #9
G113, 12, Work Offset Positioning Coordinate #10
G114, 12, Work Offset Positioning Coordinate #11
G115, 12, Work Offset Positioning Coordinate #12
G116, 12, Work Offset Positioning Coordinate #13
G117, 12, Work Offset Positioning Coordinate #14
G118, 12, Work Offset Positioning Coordinate #15
G119, 12, Work Offset Positioning Coordinate #16
G120, 12, Work Offset Positioning Coordinate #17
G121, 12, Work Offset Positioning Coordinate #18
G122, 12, Work Offset Positioning Coordinate #19
G123, 12, Work Offset Positioning Coordinate #20
G124, 12, Work Offset Positioning Coordinate #21
G125, 12, Work Offset Positioning Coordinate #22

Haas CNC Mill: G-Codes

Code, Group, Function

G126, 12, Work Offset Positioning Coordinate #23
G127, 12, Work Offset Positioning Coordinate #24
G128, 12, Work Offset Positioning Coordinate #25
G129, 12, Work Offset Positioning Coordinate #26
G136, 00, ** Automatic Work Offset Center Measurement (Option)
G141, 07, 3D + Cutter Compensation (X, Y, Z, I, J, K, D, F)
G143, 08, ** 5-Axis Tool Length Comp + (X, Y, Z, A, B, H) (Setting 15,117)
G150, 00, General Pocket Milling Routine (X, Y, P, Z, I, J, K, Q, D, R, L, S, F)
G153, 09, ** 5-Axis HSP Drill (X, Y, A, B, Z, I, J, K, Q, P, E, L, F,) (Setting 22)
G154, 09, Select Work Offset Positioning Coordinate, P1 – P99
G155, 09, ** 5-Axis Reverse Tapping Canned Cycle (X, Y, A, B, Z, J, E, L, F)
G161, 09, ** 5-Axis Drill Canned Cycle (X, Y, A, B, Z, E, L, F)
G162, 09, ** 5-AxisSpot Drill Canned Cycle (X, Y, A, B, Z, J, E, L, F)
G163, 09, ** 5-Axis Peck Drill Canned Cycle
(X, Y, A, B, Z, I, J, K, Q, P, E, L, F) (Setting 22)
G164, 09, ** 5-Axis Tapping Canned Cycle (X, Y, A, B, Z, J, E, L, F)
G165, 09, ** 5-Axis Bore In, Bore Out Canned Cycle (X, Y, A, B, Z, E, L, F)
G166, 09, ** 5-Axis Bore In, Stop, Rapid Out Canned Cycle, (Y, A, B, Z, E, L, F)
G169, 09, ** 5-Axis Bore In, Dwell, Bore Out Canned Cycle, (X, Y, A, B, Z, P, E, L, F)
G174, 00, Special Non-Vertical Rigid Tapping CCW (X, Y, Z, F)
G184, 00, Special Non-Vertical Rigid Tapping CW (X, Y, Z, F)
G187, 00, Accuracy Control High Speed Machining (E) (Setting 85)
G188, 00, Get Program From Pallet Schedule Table (PST)

Haas Row At De Anza College

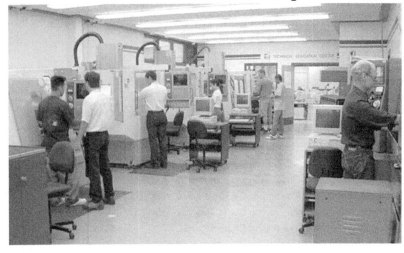

Haas CNC Mill
Letter Address Codes

A Fourth Axis Rotary Motion
Setting 30, 34, 48, 108:

The A address character is used to specify motion for the optional fourth, A-axis. It specifies an angle in degrees for the rotary axis. It is always followed by a signed number and up to three fractional decimal positions. If no decimal point is entered, the last digit is assumed to be 1/1000 degrees.

Setting 30: 4th Axis Enable
When this setting is off, it disables the 4th axis and no commands can be sent to the axis. When it is on, it is selected to one of the rotary table types to choose from in this setting. In order to change this setting the servos must be turned off (Emergency Stop in).

Setting 34: 4th Axis Diameter
This is a numeric entry. When this setting is set correctly, the surface feed rate, on the entered in the diameter for the rotary cut will be exactly the feed rate programmed into the control.

B Fifth Axis Rotary Motion
Setting 78, 79, 80, 108:

The B address character is used to specify motion for the optional fourth, B-axis. It specifies an angle in degrees for the rotary axis. It is always followed by a signed number and up to three fractional decimal positions. If no decimal point is entered, the last digit is assumed to be 1/1000 degree.

Setting 78: 5th Axis Enable
When this setting is off, it disables the 4th axis and no commands can be sent to the axis. When it is on, it is selected to one of the rotary table types to choose from in this setting. In order to change this setting the servos must be turned off (Emergency Stop in). programmed into the control.

Setting 79: 5th Axis Diameter
This is a numeric entry. When this setting is set correctly, the surface feed rate, on the entered in diameter for the rotary cut will be exactly the feed rate.

C Auxiliary External Rotary Axis
Setting 38:

The C address character is used to specify motion for the optional fourth, C-axis. It specifies an angle in degrees for the rotary axis. It is always followed by a signed number and up to three fractional decimal positions. If no decimal point is entered, the last digit is assumed to be 1/1000 degree.

Setting 38: Aux Axis Number
This is a numeric entry between 0 and 4. it is used to select the number of external auxiliary axes added to the system.

D Tool Diameter Offset Selection
Setting 40, 43, 44, 58:

The D address character is used to select the tool diameter or radius used for cutter compensation. The number following must be between 0 and 200 (100 programs on an older machine). The Dnn selects that number offset register that is in the offset display, which contains the tool diameter / radius offset amount when using cutter compensation (G41, G42). D00 will cancel cutter compensation so that the tool size is zero and it will cancel any previously defined Dnn.

Setting 40: Tool Offset Measure
Selects how the tool size is specified for cutter compensation, Radius or Diameter.

Haas CNC Mill
Letter Address Codes

E   Engraving Feed Rate Contouring Accuracy, Setting 85: The E address character is used, with G187, to select the accuracy required when cutting a corner during high speed machining operations. The range of values possible is o.0001 to 0.25 for the E code. Refer to the "Contouring Accuracy" section of your machine manual for more information.

Setting 85: Is also used to designate the same condition for Contouring Accuracy

F   Feed Rate (Setting 19, 77):

The F address character is used to select the feed rate applied to any interpolation functions, including pocket milling and canned cycles. It is either in inches per minute with four fractional positions or mm per minute with three fractional positions.

Setting 77: Allows the operator to select how the control interprets an F address code that does not contain a decimal point, (It is recommended that the programmer always use a decimal point).

G   Preparatory Functions (G Codes):

The G address character is used to specify the type of operation to occur in the block containing the G code. The G is followed by a two, or three-digit number between 0 and 187. Each G code defined in this control is part of a group of G codes. The Group O codes are non-modal; that is, they specify a function applicable to this block only and do not affect other blocks. The other groups are modal and the specification of one code in the group cancels the previous code applicable from that group. A model G code applies to all subsequent blocks so those blocks do not need to re-specify the same G code. More than one G code can be placed in a block in order to specify all of the setup conditions for an operation.

H   Tool Length Offset Selection, Setting 15: The H address character is used to select the tool length offset entry from the offsets memory. The H is followed by a two digit number between 0 and 200 (100 programs on an older machine). H0 will clear any tool length offset and Hnn will use the tool length entered in on n from the Offset Display. You must select either G43 or G44 to activate a tool length (H) offsets. The G49 command is the default condition and this command will clear any tool length offsets. A G28, M30 or pressing Reset will also cancel tool length offsets.

Setting 15: When this setting is on, a check is made to ensure that the H offset code matches the tool presently in the spindle. This check can help prevent crashes.

I   Circular Interpolation / Canned Cycle Data: The I address character is used to specify data for either canned cycles or circular motions. It is defined in inches with four fractional positions or mm with three fractional positions.

J   Circular Interpolation / Canned Cycle Data: The J address character is used to specify data for either canned cycles or circular motions. It is defined in inches with four fractional positions or mm with three fractional positions.

K   Circular Interpolation / Canned Cycle Data: The K address character is used to specify data for either canned cycles or circular motions. It is defined in inches with four fractional positions or mm with three fractional positions.

Haas CNC Mill
Letter Address Codes

L Loop Count To Repeat a Command Line:

The L address character is used to specify a repeat count for some canned cycles and auxiliary functions. It is followed by a number between 0 and 32767.

M M Code Miscellaneous Functions: The M address character is used to specify an M code. These codes are used to control miscellaneous machine functions. Note that only one M code is allowed per block in a CNC program and all M codes are performed secondary in a block.

N Number Of Block: The N address character is entirely optional. It can be used to identify or number each block of a program. It is followed by a number between 0 and 99999. The M97 functions needs to reference an N number.

O Program Number (Program Name In Parenthesis):

The O address character is used to identify a program. It is followed by a number between 0 and 99999. A program saved in memory always has a Onnnnn identification in the first block. Altering the Onnnnn in the first block causes the program to be renumbered. If you enter a program name (Program Text Name) between parentheses in the first three lines of a program, that program name will also be seen in your list of programs. You can have up to 500 program numbers (200 programs on an older machine) in your List of Programs. You can delete a program number from the LIST PROG display, by cursor selecting the program, and pressing the ERASE PROG key. You can also delete a program in the advanced editor using the menu item DELETE PROGRAM FROM LIST.

P Delay Of Time:
M98 Program Number Call:
M97 Sequence Number Call:
G103 Block Look-ahead:

The P address character is used for either a dwell time in seconds with a G04, or in canned cycles G82, G83, G86, G88, G89 and G73. When used as a dwell time, it is defined as a positive decimal value between 0.001 and 1000.0 in seconds. When P is used to search for a program number with an M98, or for a program number block in an M97. When P is used in a M97 or M98 the P value is a positive number with no decimal point up to 99999. When P is used with a G103, it defines the number of blocks the control looks-ahead in a program to execute between P1-P15.

Q Canned Cycle Optional Data:

The Q address character is used in canned cycles and is always a positive number in inches between 0.001 and 100.0.

R Circular Interpolation Canned Cycle Data, Setting 52:

The R address character is used in canned cycles and circular interpolation. It's either in inches with four fractional positions or mm with three fractional positions. It is followed by a number in inches or metric. It's usually used to define the reference plane for canned cycles.

Haas CNC Mill
Letter Address Codes

S Spindle Speed Command,
Setting 20: The S address character is used to specify the spindle speed in conjunction with M41 and M42. The S is followed by an unsigned number; between 1-99999. The S command does not turn the spindle on or off; it only sets the desired speed. If a gear change is required in order to set the commanded speed, this command will case a gear change to occur even if the spindle is stopped if spindle is running, a gear change operation will occur and the spindle will start running at the new speed.

T Tool Selection Code, Setting 15: The T address character is used to select the tool for the next tool change. The number following must be a positive number between 1 and (20) the number in Parameter 65. it dose not cause the tool change operation to occur. The Tnn may be placed in the same block that starts tool change (m06 or M16) or in any previous block.

U Auxiliary External Linear Axis: The W address character is used to specify motion for the optional external linear, W-axis. It specifies a position of motion in inches. It is always followed by a signed number and up to four fractional decimal positions. If no decimal point is entered, the last digit is assumed to be 1/10000 inches. The smallest magnitude is 0.0001 inches, the most negative value is − 8380.0000 inches, and the largest number is 8380.0000 inches.

V Auxiliary External Linear Axis: The W address character is used to specify motion for the optional external linear, W-axis. It specifies a position of motion in inches. It is always followed by a signed number and up to four fractional decimal positions. If no decimal point is entered, the last digit is assumed to be 1/10000 inches.

W Auxiliary External Linear Axis: The W address character is used to specify motion for the optional external linear, W-axis. It specifies a position of motion in inches. It is always followed by a signed number and up to four fractional decimal positions. If no decimal point is entered, the last digit is assumed to be 1/10000 inches.

X Linear X-Axis Motion, Setting 45: The X address character is used to specify motion for the X-axis. It specifies a position or distance along the X-axis. It is either in inches with four fractional positions or mm with three fractional positions. A signed number in inches or metric follows it. If no decimal point is entered, the last digit is assumed to be 1/10000 inches or 1/1000 mm.

Y Linear Y-Axis Motion, Setting 46: The Y address character is used to specify motion for the Y-axis. It specifies a position or distance along the Y-axis. It is either in inches with four fractional positions or mm with three fractional positions. A signed number in inches or metric follows it. If no decimal point is entered, the last digit is assumed to be 1/10000 inches or 1/1000 mm.

Z Linear Z-Axis Motion, Setting 47: The Z address character is used to specify motion for the Z-axis. It specifies a position or distance along the Z-axis. It is either in inches with four fractional positions or mm with three fractional positions. A signed number in inches or metric follows it. If no decimal point is entered, the last digit is assumed to be 1/10000 inches or 1/1000 mm.

Haas CNC Mill: M-Codes

All M codes are activated or cause an action to occur after everything else on a block has been completed. And only one M code is allowed per block in a program. If there is a (Setting Number) listed next to an M code, that setting will in some way relate to that M code. The following list is a summary of Haas Mill M codes.
A single asterisk (*) indicates that it's the default G code in a group.
A double asterisk (**) indicates available options.

Code, Function

M00, Program Stop (Setting 42)
M01, Optional Program Stop (Setting 17)
M02, Program End
M03, Spindle ON Clockwise (S) (Setting 144)
M04, Spindle ON Counterclockwise (S) (Setting 144)
M05, Spindle Stop
M06, Tool Change (T) (Setting 42,87,90,155)
M08, Coolant ON (Setting 32)
M09, Coolant OFF
M10**, 4-th Axis Break ON (Option)
M11**, 4-th Axis Break Release (Option)
M12**, 5-th Axis Break ON (Option)
M13**,5-th Axis Break Release (Option)
M16, Tool Change (T) (Setting 42,87,90,155)
M17**, Unclamp APC Pallet and Open APC Door
M18**, Clamp APC Pallet and Close APC Door
M19, Orient Spindle (P and R are Optional)
M21, Optional User M-Code Interface with M-Fin Signal
M22, Optional User M-Code Interface with M-Fin Signal
M23, Optional User M-Code Interface with M-Fin Signal
M24, Optional User M-Code Interface with M-Fin Signal
M25, Optional User M-Code Interface with M-Fin Signal
M26, Optional User M-Code Interface with M-Fin Signal
M27, Optional User M-Code Interface with M-Fin Signal
M28, Optional User M-Code Interface with M-Fin Signal
M30, Program End and Reset (Setting 2,39,56,83)
M31, Chip Auger Forward (Setting 114,115)
M33, Chip Auger Stop
M34, Coolant Spigot Position Down, Increment
M35, Coolant Spigot Position Up, Decrement
M36**, Pallet Part Ready (Option)
M39, Rotate Tool Turret (T) (Setting 86)
M41, Spindle Low Gear Override
M42, Spindle High Gear Override
M50**, Execute Pallet Change (Setting 121 Thru 129) (Option)

Haas CNC Mill: M-Codes

Code, Function

M51, Optional User M-Code Set
M52, Optional User M-Code Set
M53, Optional User M-Code Set
M54, Optional User M-Code Set
M55, Optional User M-Code Set
M56, Optional User M-Code Set
M57, Optional User M-Code Set
M58, Optional User M-Code Set
M59, Output Relay Set (N)
M61, Optional User M-Code Clear
M62, Optional User M-Code Clear
M63, Optional User M-Code Clear
M64, Optional User M-Code Clear
M65, Optional User M-Code Clear
M66, Optional User M-Code Clear
M67, Optional User M-Code Clear
M68, Optional User M-Code Clear
M69, Output Relay Clear (N)
M75, Set G35 or G136 Reference Point
M76, Control Display Inactive
M77, Control Display Active
M78, Alarm if Skip Signal Found
M79, Alarm if Skip Signal Not Found
M80,** Automatic Door Open (Setting 131) (Option)
M81,** Automatic Door Close (Setting 131) (Option)
M82, Tool Unclamp
M83,** Auto Air Jet ON (Option)
M84,** Auto Air Jet OFF (Option)
M86, Tool Clamp
M88, Coolant Through Spindle ON (Setting 32) (Option)
M89, Coolant Through Spindle OFF (Option)
M93, Axis Pos Capture Start (P,Q)
M94, Axis Pos Capture Stop
M95, Sleep Mode
M96, Jump if No Signal (P,Q)
M97, Local Sub-Program Call (P,L)
M98, Sub-Program Call (P,L)
M99, Sup-Program Return or Loop (P) (Setting 118)
M101,** MOM (Minimum Oil Machining) Canned Cycle Mode (I)
M102,** MOM (Minimum Oil Machining) MODE (I, J)
M103,** MOM (Minimum Oil Machining) MODE CANEL
M109, Interactive User Input (P) (Option)

Haas CNC Mill
Program Structure

A CNC part program consists of one or more blocks of commands. When viewing the program, a block is the same as a line of text. Blocks shown on the computer screen are always terminated by the " ; " (semicolon) symbol which is called an End Of Block (EOB). Blocks are made up of alphabetical address codes, which are always a letter

followed by a number. For instance, the specification to move the X-axis would be a number preceded by the X symbol, such as X2.750 the number can also have a negative value X-2.75.

Programs must begin and end with a percent (%) sign. After the first percent (%) sign with nothing else on the that line, the next line in a program must have a program number beginning with the letter O (Not Zero) and then up to five numbers (o12345) that defines that program. Those program numbers are used to identify and select a main program to be run, or as a subprogram called up by the main program. The % sign will not be seen on the control. But they must be in the program when you load a program into the Haas control. And they will be seen when you download a program from the machine. The % signs are automatically entered in for you, if you enter a program in on the Haas control.

A program may also contain a " / " symbol. The " / " symbol, sometimes called a slash or forward-slash, is used to define an optional block. If a block contains this symbol, any information that follows the slash in a program block, will be ignored when the Block Delete button is selected when running a program.

On the following page is a sample program, as it would appear on the control screen. The words in parentheses following the blocks are comments and will not be read by the Haas control. Comments must always be in parentheses so they will not be read by the Haas control. The only exception to this rule is when using G47 for text engraving.

This sample program will drill four holes and mill a two-inch hole in a four-inch square plate with X and Y at the center. The program with comment statements would appear as follows. Programs are normally written in notepad and notepad will automatically add the

(;) (semicolon) at end of block.

Haas CNC Mill
Program Structure

Program Start:

%	All programs begin and end with %
Onnnn	Program number – letter "O" and 4 numbers
(Part name & notes are optional)	(Comments must be in parentheses)
G28 G91 Z0	Tool Change position
G00 G20 G40 G49 G80 G90 M05	Insurance line
T01 M06 (tool description)	Tool # and tool change
G00 G54 Xnnn Ynnn S3000 M03	Work coordinate offset; rapid to first cutting location; spindle ON clockwise @ 3000 RPM; coolant ON.
G43 H01 Znnn M08	Activate tool length offset with register #1; rapid spindle above part; coolant on.

Continue with motion commands to machine part.

Tool Change:

G28 G91 Z0 M09	Tool Change position
G00 G20 G40 G49 G80 G90 M05	Insurance line
T02 M06 (tool description)	Tool # and tool change
G54 Xnnn Ynnn S3000 M03	Work coordinate offset; rapid to first cutting location; spindle ON clockwise @ 3000 RPM
G43 H02 Znnn M08	Activate tool length offset with register #1; rapid spindle above part; coolant on.

Continue with motion commands and tool changes to machine part.

Program End:

G00 Z.1 M09	Rapid above part; coolant Off
G28 G91 G49 Z0 M05	Spindle up to tool change position; cancel tool length offset; spindle OFF.
X0 Y0 (Optional)	Part change position, table out for easy loading and unloading.
M30	Program stop.
%	All programs begin and end with %

Note: Some programmers and/or companies prefer to activate tool length offset (G43) on the line directly following the tool number and tool change (Tnn M06)

Example T01 M06
        G43 H01 Znnn M08
        G54 Xnnn Ynnn S3000 M03 (First cutting Location)

At De Anza; TLO will **always** be turned on after position over the work-piece.

Haas CNC Mill
Program Structure

%
o10023
(Mill part program example)
(Enter Dia offset D02 at .625)
T2 M06 (5/8" Dia 2-flut end mill, tool change to tool #2)
G90 G54 G00 X-2.35 Y2.35 S1604 M03 (Abs Posit)
G43 H02 Z0.1 M08 (Tool length comp #2, Z-posit, coolant on)
G01 Z-0.625 F50. (Fast feed to depth)
G41 Y2. D02 F16. (Cutter comp left of line with Dia. comp D02)
X2.0 (Cut a 4.0" Square)
Y-2.0 (Cut a 4.0" Square)
X2.0 (Cut a 4.0" Square)
Y2.25 (Cut a 4.0" Square)
G40 X-2.3 Y2.3 (G40 cancels cutter-comp moving off part)
G00 Z1. M09 (Rapid Z1., coolant off)
G28 G91 Z0. M05 (Return Z to mach zero, spindle off)
M00 (Program stop command to check part)
T3 M06 (1/2" Dia. 90° spot drill, tool change #3)
G90 G54 G00 X-1.5 Y1.5 S1250 M03 (ABS position)
G43 H03 Z1. M08 (Tool length comp #3, Z-posit, coolant on)
G82 G99 Z-0.125 P0.2 R0.1 F10. (Spot drilling Z-.125 deep)
Y-1.5 (Spot drill #2 hole, w/optional block delete)
/X1.5 (Spot drill #3 hole, w/optional block delete)
/Y1.5 (Spot drill #4 hole, w/optional block delete)
G80 G00 Z1. M09 (Cancel canned cycle, rapid Z.1, cool off)
G28 G91 Z0. M05 (Return Z to mach zero, spindle off)
T4 M06 (1/4 Dia drill, tool change #4)
G90 G54 G00 X-1.5 Y1.5 S1400 M03 (ABS position)
G43 H04 Z1. M08 (Tool length comp #4, Z-posit, coolant on)
G83 G99 Z-0.525 Q0.5 R0.1 F12. (Peck drill to Z-.525 deep)
Y-1.5 (Peck drill #2 hole to Z-.525 deep)
/X1.5 (Peck drill #3 hole to Z-.525 deep)
/Y1.5 (Peck drill #4 hole to Z-.525 deep)
G80 G00 Z1. M09 (Cancel canned cycle, rapid Z.1, cool off)
G28 G91 Y0. Z0. M05 (Return Y & Z to mach zero)
M30 (Program stop)
% (Programs begin and end with %)

Note:
To change tools, all that is needed is an M06 even without a G28 in the previous line.
A G28 can be specified to send all axes to machine home, or it can be defined to send a specific axis home with G28 G91 Z0 and / or X0 to send just these axes specified to home position.

Haas CNC Mill
Program Subroutine

A subprogram is a separate program called up by another program. The use of subprograms can significantly reduce the amount of programming on some parts. Subroutines allow the CNC programmer to define a series of commands, which might be repeated several times in a program and, instead of repeating them many times, they can be called up when needed. A subroutine call is done with M97 or M98 and a Pnnnn. The P-code command identifies the "O" program number being used when executed with M98 or an "N" sequence number to identify the block where a local subroutine starts when executed with M87.

Local subroutines are called with an M97. This can be easier to use than the M98 because the subroutine is contained within the main program without the need to define a separate Onnnn program. With local subroutines, you define an M30 for the end of your main program portion. And after the M30 list all of your subroutines starting with a "N" sequence number to define the beginning of a local subroutine. And then end every subroutine with an M99 to send the control back to the very next line in the main program after the subroutine call.

A subroutine call from the main program calls up program blocks in a subroutine to be executed just as if they were included in the main program. Then to return back to the main program, you end a subroutine with an M99, which sends it back to the next line after the subroutine call in the main program.

Another important feature of a subroutine call is that the M97 and M98 block may also include an L-loop or repeat count. If there is an "L" number with the subroutine call, it is repeated that number of times before the main program continues with the next block.

The most common use of subroutines is in the definition of a series of holes, which may need to be center drilled, peck drilled, tapped, and/or chamfered. If a subroutine is defined that consists only of the X, Y-position of the holes, the main program can define the canned cycles, and the hole locations can be called up in the subroutine to do each of the tool operations. Thus, the X, Y-positions can be entered only once and used several times for each tool.

When a program contains duplicate patterns or sequential moves, subroutines can be used to save programming time and reduce the amount of code. A subprogram is written just like any other program except it is not designed to run by itself, but as part of the main program.

Haas CNC Mill
Program Subroutine
Executing A Subroutine

The main program, using code M98 Pnnnn, executes the subroutine. M98 says to the controller "go to program". The subprogram number is listed to identify the program being selected. A "Pnnnn" number is used in place of the "Onnnn subprogram number.
M99 is used in the last block instead of M30 and says "go to main program". The machining sequences will continue with the block after M98.

Main Program	Subprogram
O 3000	O3001
N010 . . . . ;	N010 G91
N020 . . . . ;	N020 . . . . ;
N030 M98 P3001	N030 . . . . ;
N040 . . . . ;	N040 . . . . ;
N050 . . . . ;	N050 G90
N060 M30	N060 M99

Multiple Subprograms

The main program calls subprogram A. Subprogram A executes until line N030 where it calls subprogram B. Subprogram B executes, then returns to subprogram A, which completes and returns to the main program.

Main Program	Subprogram A	Subprogram A
O 3000	O 3001	O 3000
N010. . . . . . . ;	N010 G91	N010 G91
N020. . . . . . . ;	N020. . . . . . . ;	N020. . . . . ;
N030 M98 3001	N030 M98 P3002	N030. . . . . ;
N040. . . . . . . ;	N040 . . . . ;	N040. . . . . ;
N050. . . . . . ;	N050 G90	N050 G90
N060 M30	N060 M99	N060 M99

Looping Subprograms

The "L-counter" can be used to make a subprogram execute a specified number of times before returning to the main program. M98 P3001 L2 is used in the following example to execute the subprogram two-times before returning to the main program.

Main Program	Subprogram
O 3000	O3001
N010 . . . . ;	N010 G91
N020 . . . . ;	N020 . . . . ;
N030 M98 P3001 L2	N030 . . . . ;
N040 . . . . ;	N040 . . . . ;
N050 . . . . ;	N050 G90
N060 M30	N060 M99

Haas CNC Mill
Program Subroutine
Recalling A Subprogram

A subprogram can be recalled as needed by the main program by using the M98 command.

Main Program	Subprogram
O 3000	O3001
N010 . . . . ;	N010 G91
N020 M98 P3001	N020 . . . . ;
N030 . . . . ;	N030 . . . . ;
N040 M98 P3001	N040 . . . . ;
N050 . . . . ;	N050 G90
N060 M30	N060 M99

Returning To A Line Number

By specifying a line number with the P-address with M99 on the last line of the subprogram, the control will return to the specified line number in the main program rather than the block after the M98 command. Example: M99 P050 returns controller to line N050 of the main program

Main Program	Subprogram
O 3000	O3001
N010 . . . . ;	N010 G91
N020 M98 P3001	N020 . . . . ;
N030 . . . . ;	N030 . . . . ;
N040 . . . . ;	N040 . . . . ;
N050 . . . . ;	N050 G90
N060 M30	N060 M99 P050

Notes On Subprogram Control

The subprogram call command "M98 Pnnnn" and a move command can be specified in the same block. Example:
Y-2.75 M98 P4444 is call after completing the movement in the Y-axis.

If M99 is used in the main program, the program will restart from that point. If M99 is used with P and a line number "M99 Pnnnn", the controller will go to the specified line number. This type of command is often used with the block skip (/) option. Example: / M99 Pnnnn

Since subprograms are frequently used to repeat a series of machining sequences in various locations on the part, they are often programmed in incremental G91. in the block before M99, return to main program always command a G90 absolute. Failing to do this will instruct the controller to interpret the main program values as incremental.

Subprogram A
O3001
N010 G91
N020 . . . . ;
N050 G90
N060 M99

Haas CNC Mill
Program Subroutine
Program Calling Subroutine With M98

```
%
o10106
T3 M06 (90-degree, 1/2" Dia. Spot Drill)
G90 G54 G00 X1.5 Y-0.5 (1)
S1400 M03
G43 H03 Z1. M08
G81 G99 Z-0.24 R0.1 F7. (G81 Drilling Canned Cycle)
M98 P10107 (Call Subroutine o10107)

T15 M06 (U = .368 Dia. drill)
G90 G54 G00 X1.5 Y-0.5 (1)
S2100 M03
G43 H15 Z1. M08
G83 G99 Z-0.75 Q0.2 F12.5
(G83 Peck Drill Canned Cycle)
M98 P10107 (Call Subroutine o10107)

T18 M06 (7/16-14UNC 2B tap)
G90 G54 G00 X1.5 Y-0.5
S525 (G84 will turn on the spindle, so no M03 is needed)
G43 H18 Z1. M08
G84 G99 Z-0.6 R0.1 F37.5 (G84 Tapping canned cycle)
M98 P10107 (Call Subroutine o10107)
G53 G49 Y0.
M30 (End program)
%
```

Note: See next page for subroutine being called.

Haas CNC Mill
Program Subroutine
Program Calling Subroutine With M98

%
o10107
(Subroutine, Listing all hole locations)
X0.5 Y-0.75
Y2.25
G98 X1.5 Y-2.5
G99 X3.5
X4.5 Y-2.25
Y-0.75
X3.5 Y-0.5
G80 G00 Z1.0 M09
G53 G49 Z0. M05
M99
(M99 ends subroutine and returns to main program)
(It returns to next line after the M98 subroutine call)
%

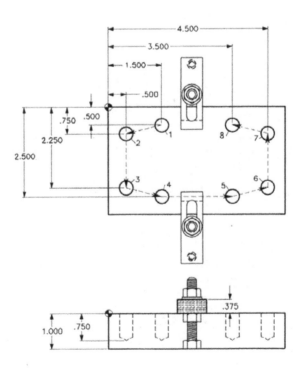

Haas CNC Mill
Circular Interpolation G02 & G03

Circular Interpolation and five pieces of information required for executing a circular interpolation commands:

1) Plane Selection, G17, Arc Parallel XY plane
1) Plane Selection, G18, Arc Parallel ZX plane
1) Plane Selection, G19, Arc parallel YZ plane

2) Arc Start Position, X,Y,Z Coordinates of Start Position

3) Rotation Direction, G02 Clockwise
3) Rotation Direction G03 Counter Clockwise

4) Arc End Position, G90 Abs. X,Y,Z of end position
4) Arc End Position, G91 Incr. X,Y,Z of end position

5) Arc Center I,J,K Dist. From Start Position in X,Y,Z
5) Arc Center Cords R (Radius)

G02 CW Circular Interpolation Motion & G03 CCW Circular Interpolation Motion

- X   Circular end point X-axis motion
- Y   Circular end point Y-axis motion
- Z   Circular end point Z-axis motion
- A   Circular end point A-axis motion
- I   X-axis distance from the start point to arc center
- J   Y-axis distance from the start point to arc center
- K   Z-axis distance from the start point to arc center
- R   Radius of the arc (If I and K are not used)
- F   Feed rate in inches (or mm) per minute

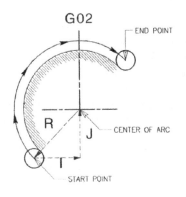 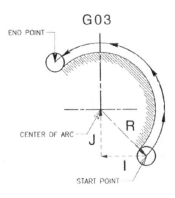

Haas CNC Mill
CW Circular Interpolation Motion G02

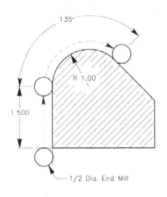

The Following line will cut an arc less than 180-degrees using a positive R Value.

G90 G54 G00 X-0.25
Y-0.25
G01 Y1.5 F12.
G02 X1.884 Y2.384 R1.25

To Generate an arc of over 180-degrees you need to specify a negative R value using R minus.

G90 G54 G00 X-0.25
Y-0.25
G01 Y1.5 F12.
G02 X1.884 Y2.384
R-1.25

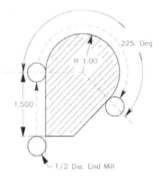

A complete 360-degree arc using an R command is not possible. To do a 360-degree arc in a G02 or G03 you need to use I and / or J to define the center of a circle for a complete 360-degree arc.

CCW Circular Interpolation Motion G03
G03 will generate counter-clockwise circular motion but is otherwise the same as G02.

Haas CNC Mill
Circular Interpolation G02 & G03

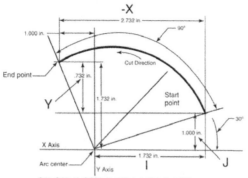

G91 G03 X -2.732 Y .732 I -1.732 J -1.000

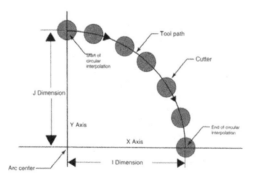

Nxx G91 G02 X 2.5000 Y -2.5000 I0 J -2.5000

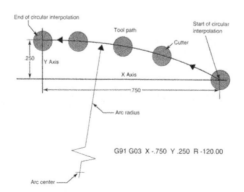

G91 G03 X -.750 Y .250 R -120.00

Haas CNC Mill
Circular Interpolation G02 & G03

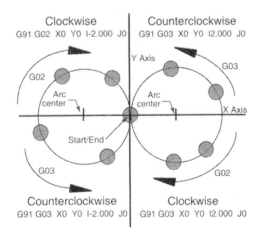

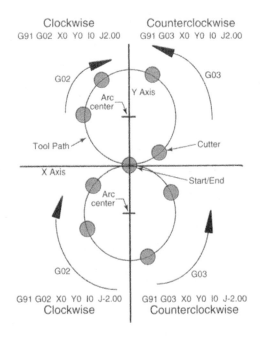

Haas CNC Mill
Circular Pocking Milling G12 & G13
One Pass "I" Only (Clockwise)

G12/G13 - "I" Only

G13 is usually preferred, for machining a counterclockwise climb cut on material

```
%
o10052
N11 (D01 Dia. offset is .500")
N12 T15 M06 (1/2" Dia. 2-flut end mill)
N13 G90 G54 G00 X2.5 Y-2.5 (X, Y Pocket center)
N14 S1910 M03
N15 G43 H15 Z0.1
N16 G13 Z-0.375 I0.5 D15 F12. (1.0 Dia. x .375 deep)
N17 G00 Z1. M09
N18 G53 G49 Y0. Z0.
N19 M30
%
```

You can feed Z-axis down with a faster or slower (N26) feed-rate than what's in the G12, G13 pocket command.

```
%
o10052
N21 (D01 Dia. offset is .500")
N22 T15 M06 (1/2" Dia. 2-flut end mill)
N23 G90 G54 G00 X2.5 Y-2.5 (X, Y Pocket center)
N24 S1910 M03
N25 G43 H15 Z0.1 M08
N26 G01 Z-0.375 F6. (Feed-rate of Z-axis before G13)
N27 G13 I0.5 D15 F10. (1.0 Dia. x .375 deep)
N28 G00 Z1. M09
N29 G53 G49 Y0. Z0.
N30 M30
%
```

Haas CNC Mill
Circular Pocking Milling G12 & G13
Multiple Passes I, K, & Q

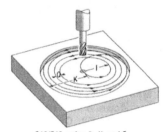

G12/G13 -using I, K, and Q

G13 is usually preferred, for machining a counterclockwise climb cut of the material
%
o10052
N31 (D02 Dia. offset is .625")
N32 T2 M06 (5/8" Dia. 2-flut end mill)
N33 G90 G54 G00 X2.5 Y-2.5 (X, Y-axis pocket center)
N34 S1520 M03
N35 G43 H02 Z0.1 M08
N36 G13 Z-0.375 I0.3 K1.5 Q0.3 D02 F9. (3.0 Dia. x .50)
N37 G00 Z1. M09
N38 G53 G49 Y0. Z0.
N39 M30
%

You can feed Z-axis down with a faster or slower (N26) feed-rate than what's in the G12, G13 pocket command.

%
o10052
N41 (D02 Dia. offset is .625")
N42 T2 M06 (5/8" Dia. 2-flut end mill)
N43 G90 G54 G00 X2.5 Y-2.5 (X, Y-axis pocket center)
N44 S1520 M03
N45 G43 H02 Z0.1 M08
N46 G01 Z-0.375 F6. (Feed-rate of Z-axis before G13)
N47 G13 I0.3 K1.5 Q0.3 D02 F9. (3.0 Dia. x .50 deep)
N48 G00 Z1. M09
N49 G53 G49 Y0. Z0.
N50 M30
%

Haas CNC Mill
Circular Pocking Milling G12 & G13
Multiple Z Passes To Depth Using
Incremental G91 And An "L" Loop Count

```
%
o10054
N51 (D02 Dia. offset is .625")
N52 T2 M06  (5/8" Dia. 2-flut end mill)
N53 G90 G54 G00 X2.5 Y-2.5 (X, Y-axis pocket center)
N54 S1520 M03
N55 G43 H02 Z0.1 M08
N56 G01 Z0. F30.
(Move down to start point to begin Increment down.)
N57 G13 G91 Z-0.375 I0.325 K2. Q0.3 D02 L4 F12.
(4.0 Dia. x 1.50 deep circular pocket)
N58 G00 Z1. M09
N59 G53 G49 Y0. Z0.
N60 M30
%
```

The above program uses G91 and a "L" count of four. This cycle is multiplied by the "L" command and will do it a total of four times at the Z-depth increment of .375 to a total depth of 1.5 inch. The G91 and "L" count can also be used for G12 and G13 "I" only line.

Since the G91 incremental Z-depth is looped together within the G13 circular pocket command, you are unable to separate the Z-axis feed-rate from the X and Y-axis feed-rate.

The feed-rate in the Z-axis will be the same as X and Y-axis.

You may want to fast feed down to the Z-surface, of where the circular pocket begins, to start incrementally step down to establish the desired depth.

Haas CNC Mill
Circular Pocket Milling G12 & G13
Milling Multiple Pockets

- X   Position to center of pocket.
- Y   Position to center of pocket.
- Z   Depth of cut or increment down.
- I   Radius first circle or the finish radius if K is not used.
- K   Radius of finished circle (If specified).
- Q   Radius step over increment (Must be used with K).
- D   Cutter Comp. number.
       (Enter cutter size into offset display register number).
- L   Loop count for repeating deeper cuts.
- F   Feed-rate in inch (mm) per min.

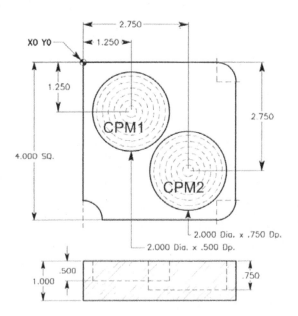

Circular Pocket Mill CPM-1 which is a 2.0 Dia. x .500 deep pocket, spiraling out to a rough 1.980 Dia. using I0.25, K0.99 and Q0.2 roughing out pocket. Then mill another circular pocket command to finish CPM-1 using 1.0 "I only" as a circular pocket finish pass.

Circular Pocket Mill CPM-2 which is a 2.0 Dia. x .750 deep pocket, incrementally stepping down 0.25 depth using an L3 count and an incremental G91 command to do 3 passes on a circular using I0.3, K1.0 and Q0.35 to machine out pocket.

(Continued on next page)

Haas CNC Mill
Circular Pocket Milling G12 & G13
Milling Multiple Pockets

Continued from previous page

```
%
o00020 (circular pocket milling exercise)
T2 M06
(T2. 5/8" Dia. 2-flut center cutting end-mill)
G90 G54 G00 X1.25 Y-1.25
S1520 M03
G43 H02 Z0.1 M08
G01 Z0.
G____ Z____ I____ K____ Q____ D____ F7.2
G____ I____ D____ F12.5
G____ Z____
X____ Y____
G01 Z0. F20.
G____ G____ Z____ I____ K____ Q____ D____ L____ F10.
G____ G____ Z____ M____
G53 G49 Z0. M05
M30
%
```

Haas CNC Mini Mills

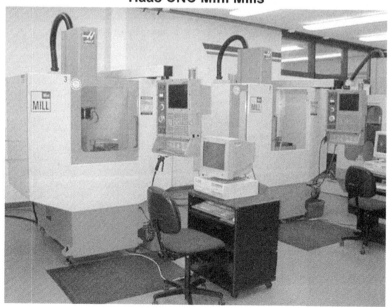

Haas CNC Mill
Circular Plane Selection G17, G18 & G19

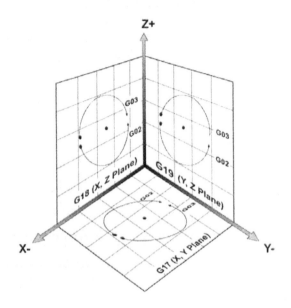

There are three G-codes for circular plane selection that are used to define the two axes, of either X, Y or Z, that you would like to do a G02 (CW) or G03 (CCW) circular motion in (XY, YZ, XZ). The plane selection is model and stays in effect for all subsequent circular interpolation moves, until you command another circular plane.

There are three G-codes used to select the circular plane; G17 for the XY plane, G18 for the XZ plane, and G19 for the YZ plane.

When machine is powered on, the default plane is G17 for the XY plane. This means G02 or G03 circular moves in the XY plane is already selected. The G17 XY-plane will always be active when you power on the machine.

To perform G02 or G03 helical motion in either G17, G18, or G19 it is possible by programming the linear axis which is not in the plane that is selected. This third axis will be moved along the specified axis in a linear manner while the other two axes are moved in a circular motion around the third axis. The speed of each axis will be controlled so that the helical rate matches the programmed feed-rate.

If cutter radius compensation G41 or G42 is selected, you can only use it in the G17 XY-circular plane. Cutter compensation for circular motions in the G18 XZ-plane or G19 YZ-plane can only be done with G141, 3D+ cutter compensation.

Haas CNC Mill
XY Circular Plane Selection G17

The G17 code is used to select the XY plane for circular motion. In this plane, circular motion is defined as clockwise for the operator looking down onto the XY table from above.

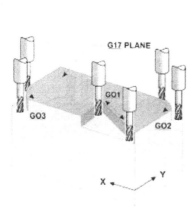
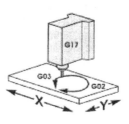

```
%
o10057
N1 T1 M06 (1/2" Dia. 4 flute end mill)
N2 G90 G54 G00 X4. Y3.25 S2600 M03
(XY start point of arc)
N3 G43 H01 Z0.1 M08
N4 G01 Z-0.375 F50.
N5 G17 G02 X5.25 Y2. R1.25 F10.
(G17 circular motion XY-plane)
(G17 is the default when you power up machine)
N6 G00 Z0.1
N7 X5. Y-.25
N8 G01 Z-0.375 F50.
N9 X3.25 Y0.8 F10.
N10 Y-.25
N11 G00 Z1.
N12 X-0.25 Y1.
N13 G01 Z-0.375 F50.
N14 G17 G03 X1. Y-0.25 R1.25 F10.
(G17 circular motion XY-plane)
N15 G00 Z1. M09
N16 G53 G49 Y0. Z0. M05
N17 M30
%
```

Haas CNC Mill
XZ Circular Plane Selection G18

The G18 code is used for doing circular motion in the XZ-plane. In these diagrams shown here, you might think the G02 and G03 arrows are incorrect. But they are correct in showing the G18 plane. You need to look at the circular direction as if you're standing at the back of the machine looking toward the spindle, for the G02 CW and G03 CCW directions in the G18 XZ circular plane.

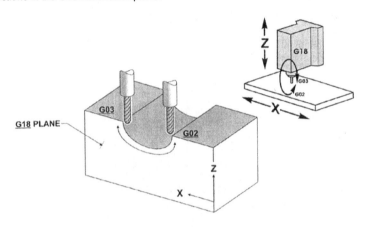

```
%
o10058
N101 T13 M06 (1/2" Dia. 2 flute ball end mill)
N102 G90 G54 G17 G00 X1.5 Y0. S2600 M03
N103 G43 H13 Z0.1 M08
N104 G01 Z0. F20.
N105 M97 P200 L80 (Local sub-call 80 times with L80)
N106 G17 G00 Z1. M09
(Switch back to G17 XY-plane when done using G18)
N107 G53 G49 Y0. Z0. M05
N108 M30
(Local sub-call N200 called by M97 P200 on line N105)
N200 G91 G01 Y-0.01
N201 G90
N202 G18 G02 X3. Z0. R0.75 F12. (G18 circular motion)
N203 G91 G01 Y-0.01
N204 G90
N205 G18 G03 X1.5 Z0. R0.75 F12. (G18 circular motion)
N206 M99
(M99 will cause the program to jump back to the next line)
(after the M97 sub-routine call in the main program)
%
```

Haas CNC Mill
YZ Circular Plane Selection G19

The G19 code is used for doing circular motion in the YZ-plane. In the G19 plane, you need to look at the circular direction as if you're standing on the right side of the machine where the machine control is, looking toward the other end, for the G02 clockwise and G03 counter-clockwise directions in the G19 YZ circular plane.

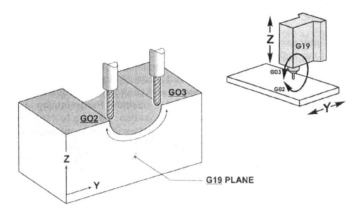

```
%
o10059
N1 T13 M06 (1/2" Dia. 2 flute ball end mill)
N2 G90 G54 G17 G00 X1.5 Y0. S2600 M03
N3 G43 H13 Z0.1 M08
N4 G01 Z0. F20.
N5 M97 P100 L80 (Local sub-call 80 times with L80)
N6 G17 G00 Z1. M09
(Switch back to G17 XY-plane when done using G19)
N7 G53 G49 Y0. Z0. M05
N8 M30
(Local sub-call N100 called by M97 P100 on line N5)
N100 G91 G01 Y-0.01
N101 G90
N102 G19 G02 X3. Z0. R0.75 F12. (G19 circular motion)
N103 G91 G01 Y-0.01
N104 G90
N105 G19 G02 X1.5 Z0. R0.75 F12. (G19 circular motion)
N106 M99
(M99 will cause the program to jump back to the next line)
(after the M97 sub-routine call in the main program)
%
```

Haas CNC Mill
Return To Reference Point, G28
Set Optional Intermediate Point

The G28 code is used to return to the machine zero position on all axes. If an X, Y, Z, or A-axis is on the same block and specifies a location, only those axes will move and return to the machines zero reference point and the movement to the machines zero reference point will be through that specified location. This point is called the intermediate point and is saved, if needed, for use in G29.

If you do not want to position through an intermediate point while specifying a specific axis to position to machine zero, then add a incremental (G91) command to this line along with a Z0, Y0 and/or X0 for the specific axis you want to send to machine zero.

This will command those axes specified to position incrementally to a zero distance as an intermediate point, and then those axes specified will then position to machine zero. Just be sure to program in an absolute (G90) command in the start-up lines for the next tool, which is usually needed for the beginning of each tool.

If no X, Y, Z, or A is specified, all axes will be moved directly to machine zero. Any auxiliary axes (B, C, . . .) are returned to there machine home after the X, Y, Z, and A-axes. G28 will not recognize any tool length offsets in this move.

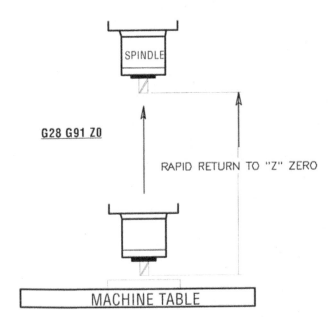

Haas CNC Mill
Cutter Compensation G40, G41 & G42

Cutter-comp is used to offset the center of the cutter, and shift it the distance of the radius, to the specified side of the programmed path. Complex part geometries having angled lines, lines tangent to arcs, and lines intersecting arcs involve substantial trigonometric computations to determine the center of the cutter. Cutter-comp involves programming the part geometry directly instead of the tool center. The cutter-comp commands are G41 cutter-comp left, G42 cutter-comp right and G40 cutter-comp cancel.

G41 Cutter-Comp Left:

G41 will select cutter-comp left; that is the tool is moved to the left of the programmed path to compensate for the radius of the tool. A Dnn must also be programmed to select the correct tool size from the Dia/Radius offset display register.

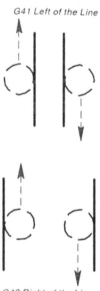

G41 Left of the Line

G42 Right of the Line

G42 Cutter-Comp Right:

G41 will select cutter-comp left; that is the tool is moved to the left of the programmed path to compensate for the radius of the tool. A Dnn must also be programmed to select the correct tool size from the Dia/Radius offset display register.

G40 Cutter-Comp Cancel:

G40 Cutter-Comp will cancel G41 or G42 cutter comp commands. A tool using cutter-comp will change from a comp-position to an uncomp-position. Programming in a D00 will also cancel cutter-comp. Cancel cutter-comp, when you're done with each milling cut series that's using Cutter-Comp.

Dnn Cutter-Comp Value:

The actual offset amount must be input in the specified tool offset display number. On the Haas Mill you have 200 tool Diameter/Radius offsets to use. Usually, you have one cutter offset for each tool, and it is best to use the same offset number as is the tool number.

Dnn Cutter-Comp Value:

n the Haas tool geometry Offset/Display page, the two columns to the right side of this display are for your cutter diameter offsets. The (Radius/Diameter) Geometry column, in the offset display, is to set your initial cutter offset value and can be defined as either a diameter or radius value, by selecting the one you want to use in Setting 40. The selection you choose will be listed at the top of the offset geometry column. The Wear column to the right of the tool (Radius/Diameter) Geometry column is for any adjustments you need to make from the initial tool Geometry offset. And these values are added together by the control and used as one value.

Haas CNC Mill
Cutter Compensation G40, G41 & G42

Two common rules:
1) If the programmed cutter path needs to Climb-Mill, which is what's usually desired on CNC machines, since the machines are rigid enough to handle this type of cut, and the conditions from this type of cut are usually preferred. And if it's a standard right handed tool, it will then be programmed with G41 cutter Left of the programmed path.

2) If the programmed cutter path needs to mill with conventional cutting, and it's a standard right-handed tool, it will then be programmed with G42 cutter Right of the Programmed path.

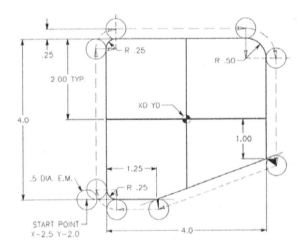

Program without cutter compensation

N103 . . . .
N104 G00 X-2.5 Y-2.0
N105 G01 Z-0.45 F50.
N106 X-2.25 F12.
N107 Y1.75
N108 G02 X-1.75 Y2.25 R0.5
N109 G01 X1.5
N110 G02 X2.25 Y1.5 R0.75

N111 G0 Y???? (Calculate point)
N112 X???? Y-2.25 (Calculate point)
N113 G01 X-1.75
N114 G02 X-2.25 Y-1.75 R0.5
N115 G01 X-2.35 Y-2.0
N116 . . . .

Haas CNC Mill
Cutter Compensation G40, G41 & G42

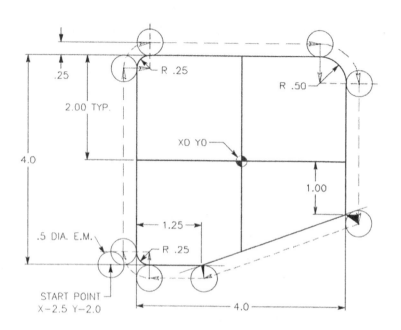

Program without cutter compensation

N103 . . . .
N104 G00 X-2.5 Y-2.0
N105 G01 Z-0.45 F50.
N106 G41 X-2.25 D01 F12.
(Turn on C/C with an X and/or Y move)
N107 Y1.75
N108 G02 X-1.75 Y2. R0.25
N109 G01 X1.5
N110 G02 X2.2 Y1.5 R0.5
N111 G1 Y-1.
N112 X-0.75 Y-2.
N113 X-1.75
N114 G02 X-2. Y-1.75 R0.25
N115 G40 G01 X-2.35
(Turn off C/C with an X and/or Y move)
N116 . . . .

Haas CNC Mill
Programming without G40, G41 & G42

When programming without cutter compensation, to the center of the cutter, a problem occurs cutting angle geometry. The cutter center must be offset to the part geometry to maintain the cutter tangency. For example, the Y-axis move from Point-A to Point-B must have the Delta-Y calculated dimension added to the .75 dimension. The X-axis move from Point-B to Point-C must have the Delta-X calculated dimension subtracted from the 1.25 dimension.

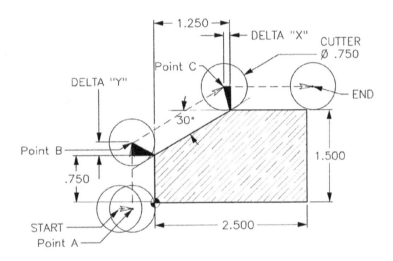

Program is to the center of the cutter manually calculating the cutter compensation of 3/4" Dia. end mill.

```
%
o10062
N1 T12 M06 (3/4" Dia. 4 flute end mill)
N2 G90 G54 G00 X-0.475 Y-0.1
(X Y position away from part , with center of tool)
N3 S1275 M03
N4 G43 H12 Z0.1 M08
N5 G01 Z-0.25 F50.
N6 X-0.375 F12.
N7 Y?.??? (Center of tool)
N8 X?.??? Y1.875 (Center of tool)
N9 X2.6
N10 G00 Z1. M09
N11 G53 G49 Y0. Z0. M05
N12 M30
%
```

Haas CNC Mill
Programming with G40, G41 & G42

When the part has been programmed using cutter compensation we are, in effect, programming with a zero diameter cutter, to the center of the tool. The tool diameter value is entered in on the Offset display page under the Radius/Diameter geometry offset column. The control will position the tool the offset amount, off of the programmed part line that is programmed so that the edge of the tool is positioning around the part geometry.

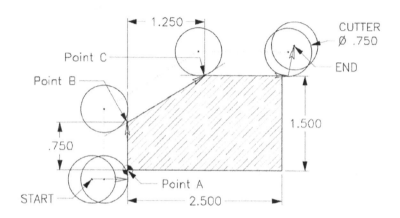

```
%
o10063
N1 T12 M06 (3/4" Dia. 4 flute end mill)
N2 G90 G54 G00 X-0.475 Y-0.1
(X Y position away from part, with center of tool)
N3 S1275 M03
N4 G43 H12 Z0.1 M08
N5 G01 Z-0.25 F50.
N6 G41 X0. D12 F12.
(Position onto part line, turning on cutter-comp)
N7 Y0.75
N8 X1.25 Y1.5
N9 X2.6
N10 G40 X2.7 Y2.
(Position off of part to the XY-center of tool)
(Turning off cutter-comp)
N11 G00 Z1. M09
N12 G53 G49 Y0. Z0. M05
N13 M30
%
```

Haas CNC Mill
Programming with G40, G41 & G42

If the cutting tool requested is not currently available, then input the available size on the offset page. Do not change the program. Cutter-Comp takes the stored value for the diameter and calculates the cutter path offset from that value. If a larger tool is going to be used, change the starting and ending positions so that the distance of the cutter is positioned half the tool diameter off the part for clearance when you lead onto and off of the part.

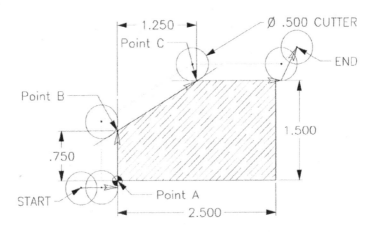

Machining part with Cutter-Comp using an undersize tool.

```
%
o10064
N1 T1 M06 (1/2" Dia. 4 flute end mill)
N2 G90 G54 G00 X-0.475 Y-0.1
(X Y position away from part, with center of tool)
N3 S1275 M03
N4 G43 H01 Z0.1 M08
N5 G01 Z-0.25 F50.
N6 G41 X0. D01 F12.
(Position onto part line, turning on cutter-comp)
N7 Y0.75
N8 X1.25 Y1.5
N9 X2.6
N10 G40 X2.7 Y2.
(Position off of part to the XY-center of tool)
(Turning off cutter-comp)
N11 G00 Z1. M09
N12 G53 G49 Y0. Z0. M05
N13 M30
%
```

Haas CNC Mill
Programming with G40, G41 & G42

Advantages Of Cutter Compensation:

1) The mathematical computations for determining a tool path are greatly simplified.

2) Because the geometry and not the tool center is programmed, the same program can be used for a variety of different cutter diameters.

3) When using cutter compensation you are then able to control and adjust for part dimensions using your cutter Diameter / Radius offsets register.

4) The same program path can be used for the roughing passes as well as finishing cuts be using different cutter offset numbers.

Some Restrictions With Cutter Compensation:

1) A cutter compensation command G41, G42, or G40 must be on the same block with an X and/or Y linear command when moving onto or off of the part using cutter-comp. You cannot turn on or off cutter-comp with a Z-axis move.

2) You can use cutter-comp in the G18 (X, Z) or G19 (Y, Z) planes using G141.

3) You cannot turn ON or OFF cutter-comp in a G02 or G03 circular move, it must be in a linear G00 or G01 straight line move.

When Activating Cutter Compensation:

1) Select a clearance point, without cutter compensation, to a start point in the X and Y-axis at least half the cutter diameter off the part before you start initiating cutter-comp.

2) Bring the Z-axis down without cutter-comp in effect.

3) Make an X and/or Y-axis move with a G41 or G42 call-out on the same line, with a diameter offset Dnn command, which has the cutter diameter value in the offset display register being used.

When Deactivating Cutter Compensation:

1) Select a clearance point X and/or Y-axis, at least half the cutter diameter off the part.

2) Do Not cancel cutter-comp on any line that is still cutting the part.

3) Cancel of cutter-comp G40, may be a one or two axis move, but you may need values entered for both X and Y-axis. This may need to be done to ensure that both axes will position to the location you want, or remain fixed and not move during the G40 cancel process. This is a programming technique that may be a programmer's preference. The code can be written with only a one-axis move, but the programmer should be aware of the results. If only one of the axes, X or Y, is on the G40 cancel command line, the control will move both axes during the cancel process. And the machine will position the axis that was not defined back to center position of that tool to the last known axis coordinate values.

Haas CNC Mill
Tool Length Offset G43, G44 & G49

G43: Activates tool length offset (TLO), sometimes referred to as tool length compensation, in a positive direction. This means the tool length offsets are added to the commanded axis positions.

G43: Selects the numeric value; which was entered into the controller register and adds it to machine reference (Zero). The compensation stays in affect until it is cancelled. The Haas cancels G43, when a tool change is made. Also G49, which is a standard G code for canceling G43, and G28 and M30 may be used.

G49: Some machines require G49 to cancel tool length compensation. The register value is removed and the spindle moves accordingly. If the spindle is positioned closer to machine zero than the register amount, the spindle will try to move beyond the machine reference position causing an over-travel alarm. Conversely, if the spindle is further from machine zero than the register amount, the spindle will stop short of the machine zero position. To eliminate these problems always use the Z0 (Z zero) command in the G49 line. This will always place the spindle at machine reference. A typical command would be:

G00 G90 G49 Z0 M05

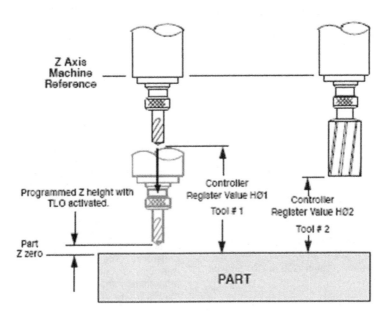

Typical Command Activating Tool Length Offset

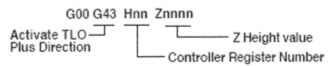

64

Haas CNC Mill
Tool Length Offset G43, G44 & G49

G43 Tool Length Compensation + (Plus)
This G43 code selects tool length compensation in a positive direction. That is; the tool length offsets are added to the commanded axis positions. An Hnn must be programmed to select the correct offset register from the offset display for the tool being used.

During the setup process, each tool point was touched-off to the part zero surface. From this position a tool length distance offset was recorded for that tool with the TOOL OFSET MESUR key. this tool length is referred to as the "Z" axis origin move to the part zero surface.

G44 Tool Length Compensation - (Minus)
This G44 code selects tool length compensation in a negative direction. That is; the tool length offsets are subtracted from the commanded axis positions. A Hnn must be programmed to select the correct entry from offsets memory.

G49 Cancels G43 / G44
This G49 code cancels tool length compensation. Putting in a H00 will also cancel tool length compensation. M30, and RESET will also cancel tool length comp.

The programmed code to position tool:

G43 H01 Z.1

TOOL CHANGER

SPINDLE

CUTTING TOOL

G43 H01 Z.1   .1

PART SURFACE

MACHINE TABLE

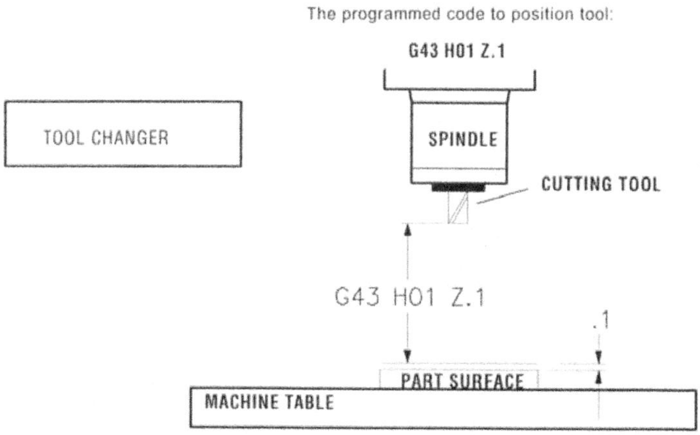

Haas CNC Mill
Thread Milling

Thread milling uses a cutter formed with the pitch of the thread to be cut. The cutters are solid carbide, fragile and expensive.

Thread milling is accomplished with helical milling. Use a standard G02 or G03 move to create the circular move in the X-Y axis and then insert a Z move on the same block corresponding to the thread pitch. This will generate one turn of the thread. If you are using a multiple tooth cutter, it will generate the rest of the thread depth. The speed of each axis will be controlled so that the helical rate matches the programmed feed-rate.

Thread milling can be used to machine larger internal or external threads. You can also thread mill, port threads, blind hole threads, metric threads and most all the special type of threads that are good to thread mill.

You can also adjust for the thread size using diameter compensation.

To do a complete 360-degree thread pass using a G02 or G03 you need to use an I and/or J to define the center of a circle for a complete thread pass. Because you cannot do a 360-degree arc using an R command.

Looping A Helical Move With A Single Point Tool

```
%
o10070
(2-12UN 3B Minor Dia = 1.9100 to 1.9198)
N31 T8 M06
N32 G90 G54 G00 X1.6 Y-1.25
N33 S1450 M03
N34 G43 H08 Z0.1 M08
N35 G01 Z-0.8 F50.
N36 G41 X2.25 D08 F10.
(Move over thread turn on C/C)
N37 G91 G03 X0. Y0. I-1. J0. Z.0833 F3. L10
(Repeat 10 times to mill thread)
N38 G90 G40 G01 X1.6 Y-1.25
(Move over thread turn on C/C)
N39 G00 Z0.1 M09
N40 G53 G49 Z0.
N41 M30
%
```

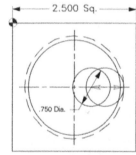

Haas CNC Mill
Thread Hob, Thread With A Single Helical Move

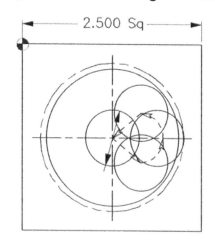

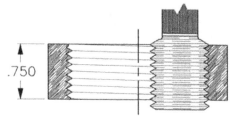

```
%
o00071
(2-12UN 3B Minor Dia = 1.9100 to 1.9198)
N51 T8 M06
N52 G90 G54 G00 X1.25 Y-1.25
N53 S1450 M03
N54 G43 H08 Z0.1 M08
N55 G01 Z-1.0 F50.
N56 G41 X1.75 Y-1.75 D08 (Position over, turn on C/C)
N57 G03 X2.25 Y-1.25 R0.5 F10. (Arcing into thread)
N58 G03 X2.25 Y-1.25 I-1. J0. Z-0.9167 F12
(To mill complete thread)
N59 G03 X1.75 Y-.75 R0.5 (Arcing off of thread)
N60 G40 G01 X1.25 Y-1.25 (Position over, turn off C/C)
N61 G00 Z0.1 M09
N62 G53 G49 Z0.
N53 M30
%
```

Haas CNC Mill
Thread Hob, Inside Diameter

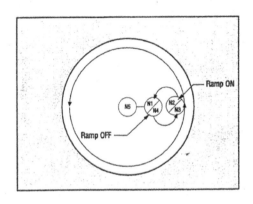
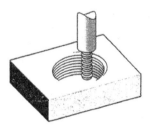

Thread milling a 1.5 diameter hole x 8 TPI

```
%
o2300
(X0, Y0 is at the center of the hole)
(Z0 is at the top of the part)
(Using .50 thick material)
G00 G90 G54 X0. Y0. S400 M03
G43 H01 Z.1 M08
Z-.6
N1 G01 G41 D01 X.175 F25 (Turn on cutter-comp)
N2 G03 X.375 R.100 F7. (Ramp on move)
N3 G03 I-.375 Z-.475
(One full revolution with Z moving up .125 inch)
N4 G03 X.175 R.100 (Ramp off move)
N5 G01 G40 X0. Y0. (Cancel cutter-comp)
G00 Z1.0 M09
G28 G91 Y0 Z0
M30
%
```

Haas CNC Mill
Thread Hob, Outside Diameter

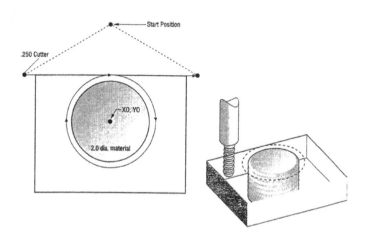

Thread milling a 2.0 diameter post x 16 TPI

%
o2400
(X0, Y0 is at the center of the post)
(Z0 is at the top of the post)
(Post height is 1.125 inch)
N1 G00 G90 G54 X0. Y2.0 S2000 M03
N2 G43 H01 Z.1 M08
N3 Z-1.0
N4 G41 D01 X-1.5 Y1.125 (Turning on cutter-comp)
N5 G01 X0.0 F15. (Linear interpolation onto the post)
N6 G02 J-1.125 Z-1.0625
(360° helical circle; negative Z move)
N7 G01X1.5 (Linear interpolation off the post)
N8 G00 G40 X0. Y2.0 (Turning off cutter-comp)
N9 Z1.0 M09
N10 G28 G91 Y0 Z0
N11 M30
%

Haas CNC Mill
Text Engraving G47

- E Plunge rate (units/minutes)
- F Engraving feed-rate (units/minutes)
- I Angle of rotation (-360.0 to +360.0), default 0°
- J Scaling factor in inches (minimum = 0.001 inches), Default is 1.0 inch.
- P 1 for Sequential Serial Number Engraving.
     0 for Literal String Engraving.
- R Return plane.
- X Start of engraving (Lower left corner of first letter).
- Y Start of engraving (Lower left corner of first letter).
- Z Depth of cut.

Note: G47 will not work with G91

This example will create the figure below.
G47 P0 X2.0 Y2.0 I45. J.5 R.05 Z-.005 F15.0 E10.0 (TEXT TO ENGRAVE)

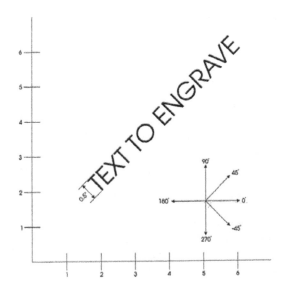

The text to engrave should be in the form of a comment on the same line as G47, with either a P1 or P0 before it.

Sequential Serial Number Engraving: this method is used to engrave numbers on a series of parts. The # symbol is used to select the number of digits in the serial number. For example: G47 P1 (####) will limit the serial number to four digits.

Haas CNC Mill
Scaling G51

This control function is optional. If you would like further information on installing this feature please call your dealer or Haas Automation for more information.

G50 Cancel Scaling:
G50 cancels scaling on all axes. Any axis scaled by a previous G51 command is no longer in effect.

G51 Scaling:
X optional center of scaling for the X-axis.

Y optional center of scaling for the Y-axis.

Z optional center of scaling for the Z-axis.

P optional scaling factor for the X-axis.
   Three-place decimal .001 to 8383.000

G51 [X...] [Y...] [Z...] [P...]

When scaling is invoked, all subsequent X, Y, Z, I, J, K, or R values pertaining to machine motion are multiplied by a scaling factor and are offset relative to a scaling center.

G51 is model and modifies appropriate positional values in the blocks following the G51 command. It dose not change or modify values in the block from which it is called. Axis X, Y, and Z are all scaled when the P code is used. If the P code is not used, the scaling factor currently in Setting 71 is used. The default scaling factor in Setting 71 is 1.0. A scaling factor of 1.0 means that no scaling is done. That is, all values are multiplied by 1.0 before being interpreted by the control.

A scaling center is always used by the control in determining the scaled position. If any scaling center is not specified in the G51 command block, then the current work coordinate position is used as the scaling center.

The following programs illustrate how scaling is performed when different scaling centers are used. All three examples call subroutine o0001which follows.

In the following illustrations for scaling.
O = Work coordinate origin.
+ = Center of scaling.

```
%
o001
F20. S500
G00 X1. Y1.
G01 X2.
Y2.
G03 X1. R0.5
G01 Y.1
G00 X0 Y0
M99
%
```

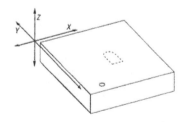

Haas CNC Mill
Scaling G51

The first example illustrates how the control uses the current coordinate location as a scaling center.

```
%
o0010
G59
G00 G90 X0 Y0 Z0
M98 P1
G51 P2. (scaling center is X0 Y0 Z0)
M30
%
```

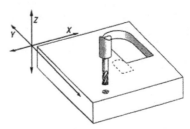

The next example specifies the center of the window as the scaling center.

```
%
o0011
G59
G00 G90 X0 Y0 Z0
M98 P1
G51 Z1.5 Y1.5 P2.
M98 P1
M30
%
```

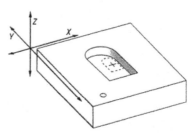

The last example illustrates how scaling can be placed at the edge of tool paths as if the part was being set against locating pins.

```
%
o0011
G59
G00 G90 X0 Y0 Z0
M98 P1
G51 X1.0 Y1.0 P2
M98 P1
M30
%
```

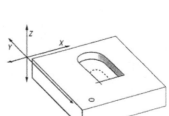

If macros are enabled, G65 arguments are not affected. Tool offsets are cutter compensation values are not affected by scaling.

The stored program is not changed by G51, so that program lines displayed by the control will not reflect actual machine positions.

Position displays will reflect the proper, scaled values.

Scaling dose not affect canned cycle Z-axis movements such as clearance planes are incremental values. The final results of scaling are rounded to the lowest fractional value of the variable being scaled.

Haas CNC Mill
Non-Modal Machine Coordinate Selection G53
Another way to return to machine Zero

This code temporarily ignores work coordinates offset and uses the machine coordinate system (machine zero). This gives you a command to move to a specific location defined from the machine zero reference point. It is non-model; so the next command block will revert back to the work coordinate offset command, that was previously active.

This G53 command may be used to send the machine home for a tool change, or to send the machine home in Y Z-axes at the end of a program, rather than using a G28. A G28 works good for sending the machine home, and most people have learned to send it home this way.

But to send it straight home with a G28, and a specific axis, you need to also define a G91 so that it positions through the intermediate point incrementally to go straight home (See G28). And then, you will usually need to be sure to switch back to a G90 absolute command, for your next move. And instead of dealing with switching in and out of absolute and incremental you could command a G53 along with a Z0, Y0, and/or X0 for the specific axis you want to send to machine zero.

Just be sure to cancel your tool length offset with a (G49) command when you define a G53 G49 Z0 to send Z-axis to machine home for a tool change.

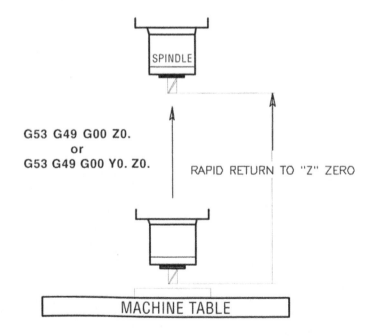

Haas CNC Mill
Rotation G68

In the example below 'a' and 'b' correspond to the axes of the current rotation plane. If G17 is the current rotation plane, then 'a' is X and 'b' is Y.

[G17 I G18 I G19] G68 [a...] [b...] [R...]
G17, G18, G19 Optional plane of rotation, default is current.

a optional center of rotation for the first axis of the selected plane.
b optional center of rotation for the second axis of the selected plane.
R optional angle of rotation specified in degrees. Three-place decimal – 360.000 to 360.000.

When rotation is invoked, all subsequent X, Y, Z, I, J, and K values are rotated through a specified rotation angle R using a center of rotation.

G68 is model and modifies appropriate positional values in the blocks following the G68 command. Values in the block containing G68 are not rotated. For subsequent blocks, only the values in the plane of rotation are rotated. Thus, if G17 is the current plane of rotation, only X and Y values are affected.

For a positive angle, the rotation is counterclockwise. If the angle of rotation – the R code – is not specified in the G68 command block, then the angle of rotation is taken for Setting 72. The default rotation angle in Setting 72 is 0.0-degrees.

A center of rotation is always used by the control to determine the positional values passed to the control after rotation. If any axis, center of rotation is not specified, then the current location of the work coordinate is used as the center of rotation.

In G90 absolute mode, the rotation angle takes on the value specified in R. When Setting 73 (G68 incremental R) is set to ON, then the rotational value can be incremented on each call to G68.

In G91 incremental mode, the rotation angle is incremented by the value in R. each G68 command block, when in G91 mode, will increment the rotation angle by the value specified in R.

Angles are modulo 360, so that when an angle is incremented past 360-degrees, the angle will become an equivalent value between 0 and 360-degrees. The rotational angle is set to zero upon cycle start, or it can be set explicitly by using a G68 block in the G90 mode.

Note: Do not change the plane of rotation while G68 is in effect.

Note: Cutter compensation should be turned on after the rotation and scaling commands are issued. Cutter compensation should also be turned off prior to turning rotation or scaling off.

G69 cancels any rotation specified previously.

Haas CNC Mill
Rotation G68

The following examples illustrate rotation-using G68.
O = work coordinate origin, + = center of rotation

```
%
o0001
(gothic window)
F20 S500
G00 X1. Y1.
G01 X2.
Y2.
G03 X1. R0.5
G01 Y1.
M99
%
```

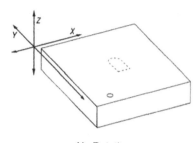

No Rotation

The first example illustrates how the control uses the current work coordinate location as a rotation center. Here, it is X0 Y0 Z0.

```
(60-degree rotation)
%
o0002
G59
G00 G90 X0 Y0 Z0
M98 P1
G90 G00 X0 Y0
G68 R60.
M98 P1
G69 G90 G00 X0 Y0
M30
%
```

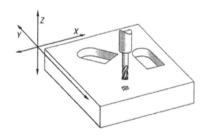

The next example specifies the center of the window as the rotation center.

```
%
o0003
G59
G00 G90 X0 Y0 Z0
M98 P1
G00 G90 X0 Y0 Z0
M68 X1.5 Y1.5 R60
M98 P1
G69 G90 G00 X0 Y0
M30
%
```

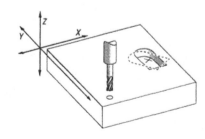

Haas CNC Mill
Rotation G68

This example shows how the G91 mode can be used to rotate pattern about a center. This is often useful for making parts that are symmetric about a regular polygon.

```
%                      %
o0004 (Program)        o0010
G59                    (Sub Program)
G00 G90 X0 Y0          G91 G68 R45.
Z0                     G90 M98 P1
M98 P0010 L8           G90 G00 X0 Y0
M30                    M99
%                      %
```

Rotation with scaling:

If scaling and rotation is used simultaneously, it is recommended that scaling is turned on prior to rotation, and that separate blocks be used. Use the following template when doing this.

```
G51 . . . . . . (scaling)
. . . . . .
G68 . . . . . . (rotation)
. . .
. . . (program)
. . .
G69 . . . . . . (rotation off)
. . . . . .
G51 . . . . . . (scaling off)
```

When rotating after scaling, any center specified as the center of rotation will be scaled. Any angle specified in the G68 block is not scald. The control applies scaling and then rotation to any block with motion commands.

Below is an example of a program that has been scaled and rotated.

```
%
o0004
G59
G00 G90 X0 Y0 Z0
M98 P1
G90 G00 X0 Y0
G51 P3.0
G68 R60.
M98 P1
G69
G51 G90 G00 X0 Y0
M30
%
```

Haas CNC Mill
Bolt Hole Circle Commands G70, G71 & G72

There are three G codes that provide commands used to do bolt hole patterns. They are G70, G71 and G72. These G codes are defined with one of the canned cycles G73, G74, G76, G77, or G81 – G89. You define the angle of the bolthole pattern, 0 to 360-degrees horizontal from three O'clock, CCW.
A minus sign will reverse angles CW.

G70 Bolt Hole Circle Command

The tool must be positioned at the center of the circle either in a previous block or in the G70 block. G70 belongs to Group Zero and thus is Non-Model. For a G70 to work correctly, a canned cycle must be active to perform the desired drill, tap or bore cycle.

I = Radius of the bolt hole circle.
J = Starting angle from three o'clock, 0 to 360-degrees.
L = No. of evenly spaced holes around bolt hole circle.

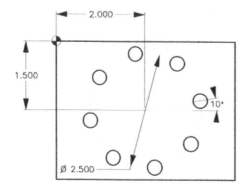

%
o00099 (Bolt Hole Circle)
N1 T15 M06 (3/8 Dia. drill)
N2 G90 G54 G00 X2. Y-1.5
(center of bolt circle)
N3 S1620 M03
N4 G43 H15 Z1. M08
N5 G81 G99 Z-0.45 R0.1 F8. L0
N6 G70 I1.25 J10. L8
N7 G80 G00 Z1. M09
N8 G53 G49 Z0. M05
N9 M30
%

Having L0 on line N5 will cause machine to not do this command until the control reads the next line, so as not to drill a hole in the center of bolt circle. Or you can combine N5 and N6 together, minus the L0, to also not drill a hole in the center.

Haas CNC Mill
Bolt Hole Arc Command G71

The tool must be positioned at the center of the circle either in a previous block or in the G70 block. G70 belongs to Group Zero and thus is Non-Model. For a G70 to work correctly, a canned cycle must be active to perform the desired drill, tap or bore cycle. Using a K minus value (K-30.0) defines positional around the bolt hole arc CW starting from the J angle.

I = Radius of the bolt hole arc.
J = Starting angle from three o'clock, 0 to 360-degrees.
K = Angle spacing between holes (+ or -)
L = No. of evenly spaced holes around bolt hole circle.

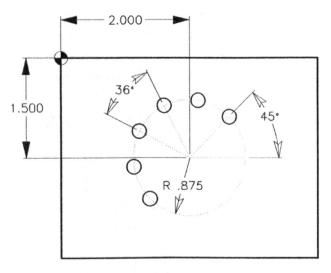

```
%
o00100 (Bolt Hole Circle)
N1 T15 M06 (3/8 Dia. drill)
N2 G90 G54 G00 X2. Y-1.5
(center of bolt circle)
N3 S1450 M03
N4 G43 H15 Z1. M08
N5 G81 G99 Z-0.45 R0.1 F8. L0
N6 G71 I0.875 J45. L6
N7 G80 G00 Z1. M09
N8 G53 G49 Z0. M05
N9 M30
%
```

Having L0 on line N5 will cause machine to not do this command until the control reads the next line, so as not to drill a hole in the center of bolt circle. Or you can combine N5 and N6 together, minus the L0, to also not drill a hole in the center.

Haas CNC Mill
Bolt Holes Along An Angle G72

The G72 code drills L holes in a straight line at the specified angle. G72 belongs to Group Zero and thus is non-model. For a G72 to work correctly. A canned cycle must be active to perform the desired drill, tap or bore cycle.

I = Distance between bolt holes along an angle.
J = Angle of hole from three o'clock, 0 to 360-degrees.
L = Number of evenly spaced holes along an angle.

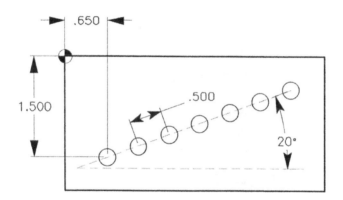

```
%
o00101 (Bolt Holes Along An Angle)
N1 T16 M06 (1/2 Dia. Drill)
N2 G90 G54 G00 X0.65 Y-1.5
(Start position of bolt holes along an angle)
N3 S1450 M03
N4 G43 H16 Z1. M08
N5 G81 G99 Z-0.45 R0.1 G72 I0.5 J20.0 L7 F8.
N6 G80 G00 Z1. M09
N7 G53 G49 Z0. M05
N8 M30
%
```

Haas CNC Mill
Bolt Hole Pattern Repeat Command " L "

The L repeat command directs the controller to repeat a motion or operation a specified number of times. It is most commonly used with canned cycles where a number of equally spaced moves are required. The L command is not modal, it affects only the information in the same block. Model commands from a previous block, such as a canned cycle, are still in affect. A number following the L command specifies the number of repeats. An axis move used with the L command (ie: Y-2.000 L5) must be used in incremental mode.

Example: Of one way the L command can be used

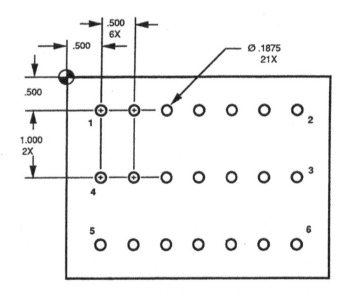

G90 X0 Y0 (Absolute mode, spindle at program zero)
G81 G99 X.500 Y-.500 Z-.300 R.100 F10. (Drill hole #1)
G91 X.5 L6 (Repeat drill cycle, incremental, 6 times to #2)
Y-1.00 (Incremental move to #3)
X-.500 L6 (Repeat drill cycle, incremental, 6 times to #4)
Y-1.000 (Incremental move to #5)
X.5 L6 (Repeat drill cycle, incremental, 6 times to #6)
G90 G80 X0 Y0
(Absolute mode, cancel canned cycle, return to zero)

Haas CNC Mill
Canned Cycles

G73, High speed peck drilling
Z Drilling Operation: Intermittent feed in
Operation At The End of Hole:
Dwell (optional)
Z-axis Retraction: Rapid out

G74, Left hand tapping
Z Drilling Operation:
Spindle CCW feed in
Operation At The End of Hole:
Spindle reverse CW
Z-axis Retraction: Feed out

G76, Fine boring
Z Drilling Operation: Feed in
Operation At The End of Hole: Dwell, spindle stop, orient, shift
Z-axis Retraction: Rapid out

G77, Back boring
Z Drilling Operation: Spindle stop, shift, rapid in, shift, feed up
Operation At The End of Hole: Orient spindle, shift
Z-axis Retraction: Rapid out

G80 Canned Cycles Cancel:
The G80 code is used to cancel a canned cycle. This G code is modal in that it deactivates all canned cycles G73, G74, G76, G77, OR G81 G89 until a new one is selected. G00 or G01 code will also cancel any active canned cycle

G81, Drilling
Z Drilling Operation: Feed in
Operation At The End of Hole: None
Z-axis Retraction: Rapid out

G82, Spot drilling
Z Drilling Operation: Feed in
Operation At The End of Hole: Dwell
Z-axis Retraction: Rapid out

G83, Application: Peck drill cycle
Z Drilling Operation: Intermittent feed in
Operation At The End of Hole:
Dwell (optional)
Z-axis Retraction: Rapid out

G84, Application: Tapping cycle
Z Drilling Operation: Spindle CW feed in
Operation At The End of Hole:
Spindle reverse CCW
Z-axis Retraction: Feed out

G85, Boring cycle
Z Drilling Operation: Feed in
Operation At The End of Hole: None
Z-axis Retraction: Feed Out

G86, Boring cycle
Z Drilling Operation: Feed in
Operation At The End of Hole:
Spindle stop, orient
Z-axis Retraction: Rapid out

G87, Boring cycle
Z Drilling Operation: Feed in
Operation At The End of Hole:
Spindle stop, orient
Z-axis Retraction: Manual jog out

G88, Boring cycle
Z Drilling Operation: Feed in
Operation At The End of Hole:
Dwell, spindle stop, orient
Z-axis Retraction: Manual jog out

G89, Boring cycle
Z Drilling Operation: Feed in
Operation At The End of Hole: Dwell
Z-axis Retraction: Rapid out

Haas CNC Mill
G73 High Speed Peck Drill, Canned Cycle, using Q

- X   Rapid X-axis location (optional)
- Y   Rapid Y-axis location (optional)
- Z   Z-depth (Feed to Z-depth starting from R plane
- Q   Pecking equal incremental depth amount
      (If I, J and K are not used) (optional)
- I   Size of first peck depth (If Q is not used) (optional)
- J   Amount reducing each peck after first peck depth
      (If Q is not used) (optional)
- K   Minimum peck depth (If Q is not used) (optional)
- P   Dwell time at Z-depth (optional)
- R   R-Plane (Rapid point to start feeding)
- F   Feed-rate in inches (mm) per minute

This G code is modal so that it is activated every X and/or Y-axis move, and it will rapid to that position and then cause this canned cycle to be executed again, until it's canceled. The depth of each peck in this cycle will be the amount defined with Q or using I, J, & K or K & Q. the tool will pull back after each peck and then back in for the next peck until Z-depth is reached. Is reached. This cycle is a high-speed peck cycle were the retract distance it pulls back after each peck is set by Setting 22. Use G98 and G99 for the Z position clearance location for positioning between holes.

```
%
o10093
N1 T10 M06  (7/8" Dia. Insert Drill)
N2 G90 G54 G00 X0.625 Y0.625
N3 S1450 M03
N4 G43 H10 Z1. M08
N5 G73 G99 Z-2.15 Q0.1 R0.1 F9.
N6 X1.375 Y1.375
N7 G80 G00 Z1. M09
N8 G53 G49 Z0. M05
N9 M30
%
```

Haas CNC Mill
G73 High Speed Peck Drill, Canned Cycle, using Q

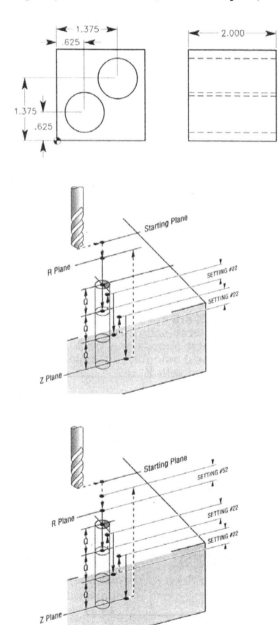

Haas CNC Mill
G73 High Speed Peck Drill, Canned Cycle, using I, J, K

- X   Rapid X-axis location (optional)
- Y   Rapid Y-axis location (optional)
- Z   Z-depth (Feed to Z-depth starting from R plane)
- Q   Pecking equal incremental depth amount
       (If I, J and K are not used) (optional)
- I   Size of first peck depth (If Q is not used) (optional)
- J   Amount reducing each peck after first peck depth
       (If Q is not used) (optional)
- K   Minimum peck depth (If Q is not used) (optional)
- P   Dwell time at Z-depth (optional)
- R   R-Plane (Rapid point to start feeding)
- F   Feed-rate in inches (mm) per minute

This G code is modal so that it is activated every X and/or Y-axis move, and it will rapid to that position and then cause this canned cycle to be executed again, until it's canceled. The depth of each peck in this cycle will be the amount defined with Q or using I, J, & K or K & Q. the tool will pull back after each peck and then back in for the next peck until Z-depth is reached. Is reached. This cycle is a high-speed peck cycle were the retract distance it pulls back after each peck is set by Setting 22. Use G98 and G99 for the Z position clearance location for positioning between holes.

If I, J, and K are specified, a different operating mode is selected. The first peck will be in by I, each succeeding peck will be reduced by the J amount, with a minimum peck being defined with K.

```
%
o10094
N1 T10 M06 (7/8" Dia. Insert Drill)
N2 G90 G54 G00 X0.625 Y0.625
N3 S1450 M03
N4 G43 H10 Z1. M08
N5 G73 G99 Z-2.15 I0.4 J0.05 K0.1 R0.1 F9.
N6 X1.375 Y1.375
N7 G80 G00 Z1. M09
N8 G53 G49 Z0. M05
N9 M30
%
```

Haas CNC Mill
G73 High Speed Peck Drill, Canned Cycle, using I, J, K

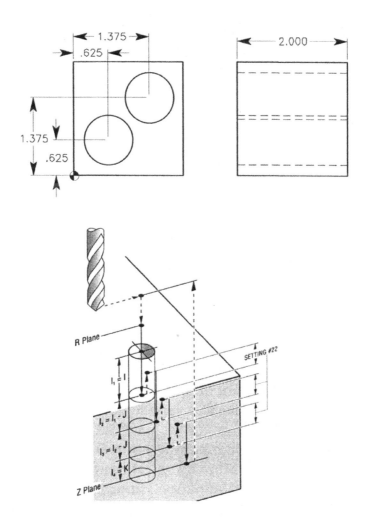

Haas CNC Mill
G73 High Speed Peck Drill, Canned Cycle, using K & Q

- X   Rapid X-axis location (optional)
- Y   Rapid Y-axis location (optional)
- Z   Z-depth (Feed to Z-depth starting from R plane
- Q   Pecking equal incremental depth amount
      (If I, J and K are not used) (optional)
- I   Size of first peck depth (If Q is not used) (optional)
- J   Amount reducing each peck after first peck depth
      (If Q is not used) (optional)
- K   Minimum peck depth (If Q is not used) (optional)
- P   Dwell time at Z-depth (optional)
- R   R-Plane (Rapid point to start feeding)
- F   Feed-rate in inches (mm) per minute

This G code is modal so that it is activated every X and/or Y-axis move, and it will rapid to that position and then cause this canned cycle to be executed again, until it's canceled. The depth of each peck in this cycle will be the amount defined with Q or using I, J, & K or K & Q. the tool will pull back after each peck and then back in for the next peck until Z-depth is reached. Is reached. This cycle is a high-speed peck cycle were the retract distance it pulls back after each peck is set by Setting 22. Use G98 and G99 for the Z position clearance location for positioning between holes.

If K and Q commands are specified together with a G73, a different operating mode is selected in this cycle. After a number of pecks of Q distance down into the part totals up to the K amount, and then multiples of K thereafter, the tool will then return to the R-plane. This allows much faster drilling then a G83, but still returns to the R-plane occasionally to clear chips.

```
%
o10095
N1 T10 M06 (7/8" Dia. Insert Drill)
N2 G90 G54 G00 X0.625 Y0.625
N3 S1450 M03
N4 G43 H10 Z1. M08
N5 G73 G99 Z-2.15 K1. Q0.2 R0.1 F9.
N6 X1.375 Y1.375
N7 G80 G00 Z1. M09
N8 G53 G49 Z0. M05
N9 M30
%
```

Haas CNC Mill
G73 High Speed Peck Drill Canned Cycle, using K & Q

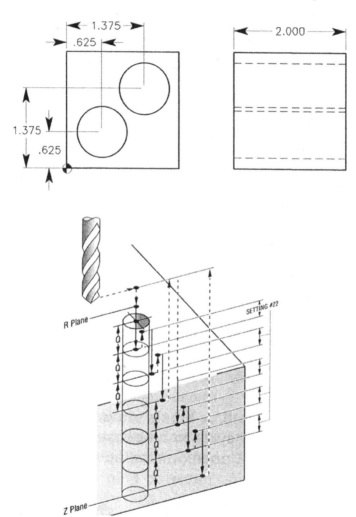

Haas CNC Mill
G74 Reverse (Left Hand) Tapping, Canned Cycle

- X Rapid X-axis location. (optional)
- Y Rapid Y-axis location. (optional)
- Z Z-depth. (Tapping Z-depth starting from R-plane)
- J Tapping retract speed (Rev. 10.13 and above). (optional)
- R R-plane. (Rapid point to start feeding)
- F Feed-Rate in inches (mm) per minute.

This G code is model. Use G98 and G99 for the Z position clearance location. You don't need to start the spindle with a M03 for a tap that's using G74 because this cycle will turn on the spindle for you automatically and it will do it quicker.

With rigid tapping, the ratio between federate and spindle speed must be calculated for the pitch being cut.

The calculation is (1/TPI) x RPM = Tapping feed-rate.
Use the Haas calculator for the speed and feed numbers.

```
%
o10083 (G74 Left hand tapping cycle)
N1 T18 M06 (7/16-14 Left handed tap)
N2 G90 G54 G00 X0.625 Y0.625
N3 S490
(You don't need M03, G84 turns on the spindle for you.)
N4 G43 H18 Z1. M08
N5 G74 G99 Z-0.65 R0.1 J3 F35.
N6 X-0.625 Y-0.625
N7 G80 G00 Z1. M09
N8 G53 G49 Z0. M05
N9 M30
%
```

Haas CNC Mill
G74 Reverse (Left Hand) Tapping, Canned Cycle

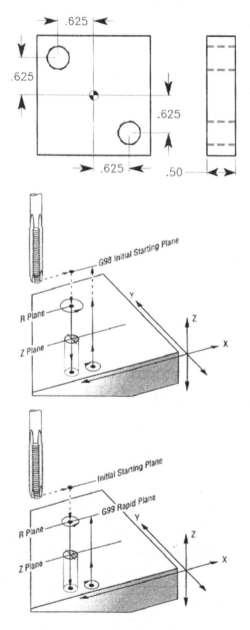

Haas CNC Mill
G76 Bore In, Stop, Shift, Rapid Out, Canned Cycle

- X   Rapid X-axis location (optional)
- Y   Rapid Y-axis location (optional)
- Z   Z-depth (Feed to Z-depth starting from R plane
- P   Dwell time at Z-depth (optional)
- Q   Shift value, always incremental (If I, or J are not used) (optional)
- I   Shift value X-axis (If Q is not used) (optional)
- J   Shift value Y-axis (If Q is not used) (optional)
- R   R-Plane (Rapid point to start feeding)
- F   Feed-rate in inches (mm) per minute

This G code is modal so that it is activated every X and/or Y-axis move, and it will rapid to that position and then cause this canned cycle to be executed again, until it's canceled. G98 and G99 for the Z position clearance location for positioning between holes.

```
%
o10096
N1 T19 M06 (Boring Bar)
N2 G90 G54 G00 X1.0 Y-1.25
N3 S1450 M03
N4 G43 H19 Z1. M08
N5 G76 G99 Z-0.55 P0.2 I-0.01 R0.1 F4.5
N6 G80 G00 Z1. M09
N8 G53 G49 Z0. M05
N9 M30
%
```

Haas CNC VF-0 Mill

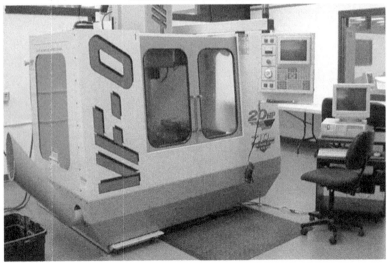

Haas CNC Mill
G76 Bore In, Stop, Shift, Rapid Out, Canned Cycle

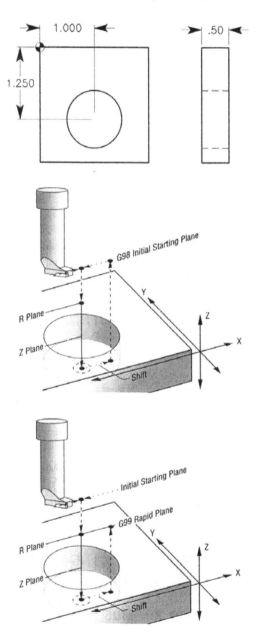

Haas CNC Mill
G76 Bore In, Stop, Shift, Rapid Out, Canned Cycle
G77 Back Bore, Canned Cycle

- X Rapid X-axis location (optional)
- Y Rapid Y-axis location (optional)
- Z Z-depth (Feed to Z-depth starting from R plane
- Q Shift value, always incremental (If I, or J are not used) (optional)
- I Shift value X-axis (If Q is not used) (optional)
- J Shift value Y-axis (If Q is not used) (optional)
- R R-Plane (Rapid point to start feeding)
- F Feed-rate in inches (mm) per minute

This G code is modal so that it is activated every X and/or Y-axis move, and it will rapid to that position and then cause this canned cycle to be executed again, until it's canceled. G98 and G99 for the Z position clearance location for positioning between holes.

This cycle will shift the X and/or Y-axis prior to and after cutting in order to clear the tool while entering and exiting the part. The Q-value shift direction is set with Setting 27. If Q is not specified, the optional I and J-values can be used to determine shift direction and distance. The tool will rapid down to the R clearance depth and feed up to the Z command depth for back counterbore.

```
%
o10097
N1 T19 M06 (Back Boring Bar)
N2 G90 G54 G00 X1.25 Y-0.75
N3 S1450 M03
N4 G43 H19 Z1. M08
N5 G77 G99 Z-0.4 R-0.55 Q0.12 F4.5
N6 G80 G00 Z1. M09
N8 G53 G49 Z0. M05
N9 M30
%
```

Haas CNC Mill
G77 Back Bore, Canned Cycle

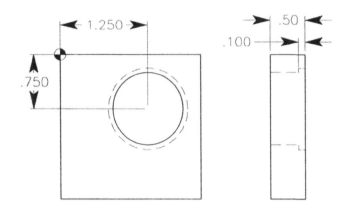

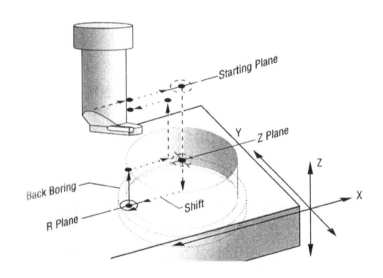

Haas CNC Mill
G81 Drill, Canned Cycle

X   Rapid X-axis location. (optional)
Y   Rapid Y-axis location. (optional)
R   Z-depth. (Feed to Z-depth starting from R-plane)
F   Feed-Rate in inches (mm) per minute.

This G code is modal so that it is activated every X and/or Y-axis move, and it will rapid to that position and then cause this canned cycle to be executed again, until it's canceled. Use G98 and G99 for the Z position clearance location for positioning between holes.

```
%
o10075 (G81 Drilling cycle)
N1 T16 M06 (1/2" Dia. Drill)
N2 G90 G54 G00 X0.5 Y-0.5
N3 S1450 M03
N4 G43 H16 Z1. M08
N5 G81 G99 Z-0.375 R0.1 F9.
N6 X1.5
N7 Y-1.5
N8 X0.5
N9 G80 G00 Z1. M09
N10 G53 G49 Z0. M05
N11 M30
%
```

Haas CNC VF-0 Mill

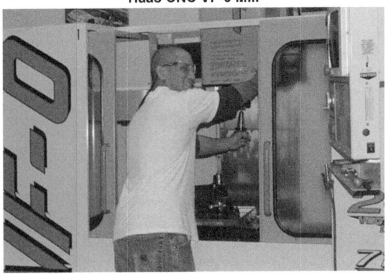

Haas CNC Mill
G81 Drill, Canned Cycle

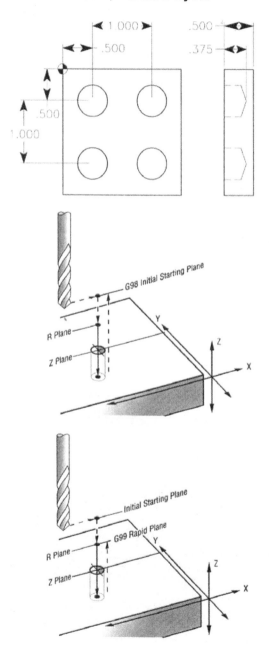

Haas CNC Mill
G82 Spot Drill & Counterbore, Canned Cycle

- X Rapid X-axis location. (optional)
- Y Rapid Y-axis location. (optional)
- Z Z-depth. (Feed to Z-depth starting from R-plane)
- P Dwell time at Z-depth.
- R R-plane. (Rapid point to start feeding)
- F Feed-Rate in inches (mm) per minute.

This G code is modal so that it is activated every X and/or Y-axis move, and it will rapid to that position and then cause this canned cycle to be executed again, until it's canceled. A dwell in seconds/milliseconds is caused at the bottom of each Z-depth in this cycle which is defined with P. Use G98 and G99 for the Z position clearance location for positioning between holes.

```
%
o10076 (G82 Drill, dwell cycle)
N1 T11 M06 (1/2" Dia. 2-flute end-mill)
N2 G90 G54 G00 X0.5 Y-0.5
N3 S1200 M03
N4 G43 H11 Z1. M08
N5 G82 G99 Z-0.375 P0.5 R0.1 F7.5
N6 X1.5
N7 Y-1.5
N8 X0.5
N9 G80 G00 Z1. M09
N10 G53 G49 Z0. M05
N11 M30
%
```

Haas CNC Horizontal Mill EC-300

Haas CNC Mill
G82 Spot Drill & Counterbore, Canned Cycle

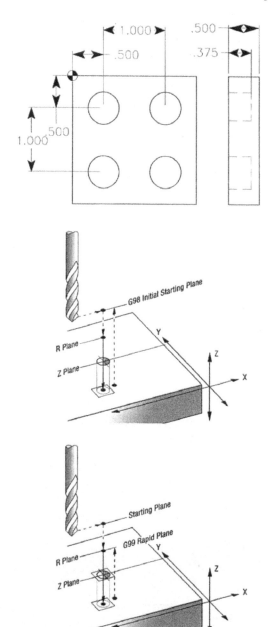

Haas CNC Mill
G83 Normal Peck Drilling, using Q

- X   Rapid X-axis location. (optional)
- Y   Rapid Y-axis location. (optional)
- Z   Z-depth. (Feed to Z-depth starting from R-plane)
- Q   Pecking equal incremental depth amount. (optional)
- P   Dwell time at Z-depth.
- R   R-plane. (Rapid point to start feeding)
- F   Feed-Rate in inches (mm) per minute.

This G code is modal so that it is activated every X and/or Y-axis move, and it will rapid to that position and then cause this canned cycle to be executed again, until it's canceled. And the depth for each peck in this cycle will be the amount defined with Q. then the tool will rapid up to the R-plane after each peck and then back in for the next peck until Z-depth is reached. Use G98 and G99 for the Z position clearance location for positioning between holes.

```
%
o10078 (G83 Deep hole peck drilling using Q)
N1 T10 M06 (7/8" Dia. drill)
N2 G90 G54 G00 X0.625 Y-0.625
N3 S1050 M03
N4 G43 H10 Z1. M08
N5 G83 G99 Z-2.3 Q0.5 R0.1 F8.
N6 X1.375 Y1.375
N7 G80 G00 Z1. M09
N8 G53 G49 Z0. M05
N09 M30
%
```

Haas CNC Horizontal Mill EC-300

Haas CNC Mill
G83 Normal Peck Drilling

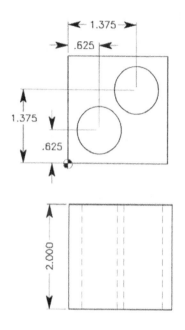

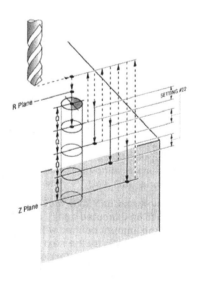

Haas CNC Mill
G83 Peck Drilling With I, J, & K options

- X   Rapid X-axis location. (optional)
- Y   Rapid Y-axis location. (optional)
- Z   Z-depth. (Feed to Z-depth starting from R-plane)
- I   Size of first peck depth (optional)
- J   Amount reducing each peck after first peck depth. (optional)
- K   Minimum peck depth. (optional)
- P   Dwell time at Z-depth.
- R   R-plane. (Rapid point to start feeding)
- F   Feed-Rate in inches (mm) per minute.

This G code is modal so that it is activated every X and/or Y-axis move, and it will rapid to that position and then cause this canned cycle to be executed again, until it's canceled. If I, J, and K are specified, a different operating mode is selected. The first pass will cut in by I, each succeeding cut will be reduced by amount J, and the minimum cutting depth is K. Use G98 and G99 for the Z position clearance location for positioning between holes.

```
%
o10079 (G83 Deep hole peck drilling using I, J, & K)
N1 T16 M06 (1/2" Dia. drill)
N2 G90 G54 G00 X0.625 Y-0.625
N3 S1833 M03
N4 G43 H16 Z1. M08
N5 G83 G99 Z-2.18 I0.5 J0.1 K0.2 R0.1 F9.
N6 X1.375 Y1.375
N7 G80 G00 Z1. M09
N8 G53 G49 Z0. M05
N09 M30
%
```

Haas CNC Mill
G83 Peck Drilling With I, J, & K options

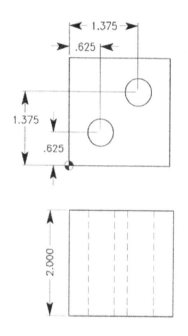

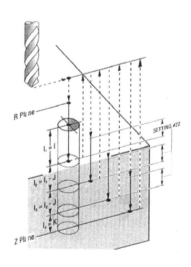

Haas CNC Mill
G84 Tapping (Right Hand) Canned Cycle

- X   Rapid X-axis location. (optional)
- Y   Rapid Y-axis location. (optional)
- Z   Z-depth. (Tapping Z-depth starting from R-plane)
- J   Tapping retract speed (Rev. 10.13 and above) (optional)
- R   R-plane. (Rapid point to start feeding)
- F   Feed-Rate in inches (mm) per minute.

This G code is model. Use G98 and G99 for the Z position clearance location. You don't need to start the spindle with a M03 for a tap that's using G84 because this cycle will turn on the spindle for you automatically and it will do it quicker.

With rigid tapping, the ratio between federate and spindle speed must be calculated for the pitch being cut.

The calculation is (1/TPI) x RPM = Tapping feed-rate.
Use the Haas calculator for the speed and feed numbers.

```
%
o10082 (G84 Right hand tapping cycle)
N1 T18 M06  (7/16-14 Right handed tap)
N2 G90 G54 G00 X0.625 Y0.625
N3 S500
(You don't need M03; G84 turns on the spindle for you.)
N4 G43 H18 Z1. M08
N5 G84 G99 Z-0.65 R0.1 J3 F35.7143
N6 X-0.625 Y-0.625
N7 G80 G00 Z1. M09
N8 G49 Z0. M05
N9 M30
%
```

Haas CNC Mill
G84 Tapping (Right Hand) Canned Cycle

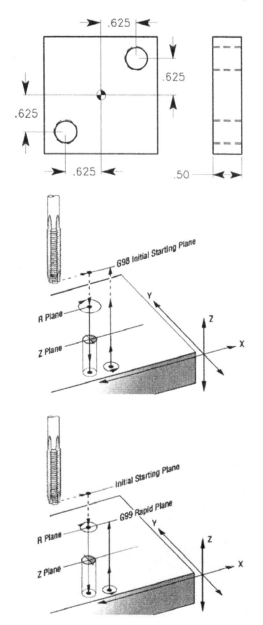

Haas CNC Mill
G85 Bore In, Bore Out Canned Cycle

- X Rapid X-axis location. (optional)
- Y Rapid Y-axis location. (optional)
- Z Z-depth. (Feed to Z-depth starting from R-plane)
- R R-plane. (Rapid point to start feeding)
- F Feed-Rate in inches (mm) per minute.

This G code is modal so that it is activated every X and/or Y-axis move, and it will rapid to that position and then cause this canned cycle to be executed again, until it's canceled. Use G98 and G99 for the Z position clearance location for positioning between holes.

```
%
o10084 (G85 Bore in, bore out cycle)
N1 T19 M06 (Boring bar)
N2 G90 G54 G00 X0.00 X0.5 Y0.5
N3 S1450 M03
N4 G43 H19 Z1. M08
N5 G85 G99 Z-.54 R0.1 F4.5
N6 X-0.5
N7 Y-0.5
N8 G80 G00 Z1. M09
N9 G53 G49 Z0. M05
N10 M30
%
```

Haas CNC Mini Mill

Haas CNC VF-0 Mill

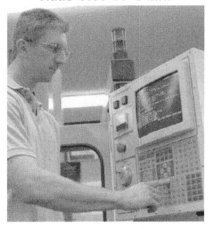

Haas CNC Mill
G85 Bore In, Bore Out Canned Cycle

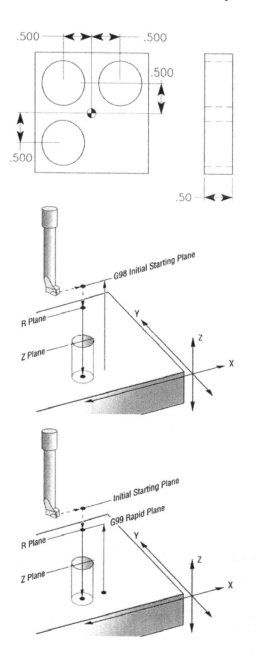

Haas CNC Mill
G86 Bore In, Stop, Rapid Out Canned Cycle

X   Rapid X-axis location. (optional)
Y   Rapid Y-axis location. (optional)
Z   Z-depth. (Feed to Z-depth starting from R-plane)
R   R-plane. (Rapid point to start feeding)
F   Feed-Rate in inches (mm) per minute.

This G code is modal so that it is activated every X and/or Y-axis move, and it will rapid to that position and then cause this canned cycle to be executed again, until it's canceled. Use G98 and G99 for the Z position clearance location between holes.

```
%
o10085 (G86 Bore in, stop, rapid out cycle)
N1 T19 M06 (Boring bar)
N2 G90 G54 G00 X0.00 X0.5 Y0.5
N3 S1450 M03
N4 G43 H19 Z1. M08
N5 G86 G99 Z-.54 R0.1 F4.5
N6 X-0.5
N7 Y-0.5
N8 G80 G00 Z1. M09
N9 G53 G49 Z0. M05
N10 M30
%
```

Mike Appio, Assisting Student

Haas CNC Mill
G86 Bore In, Stop, Rapid Out Canned Cycle

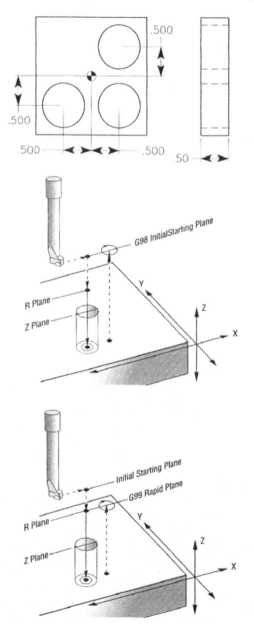

Haas CNC Mill
G87 Bore In, Manual Retract Canned Cycle

X   Rapid X-axis location. (optional)
Y   Rapid Y-axis location. (optional)
Z   Z-depth. (Feed to Z-depth starting from R-plane)
R   R-plane. (Rapid point to start feeding)
F   Feed-Rate in inches (mm) per minute.

This G code is modal so that it is activated every X and/or Y-axis move, and it will rapid to that position and then cause this canned cycle to be executed again, until it's canceled. Use G98 and G99 for the Z position clearance location between holes.

```
%
o10086 (G87 Bore in, manual retract cycle)
N1 T19 M06 (Boring bar)
N2 G90 G54 G00 X0.00 X0.5 Y0.5
N3 S1450 M03
N4 G43 H19 Z1. M08
N5 G87 G99 Z-0.54 R0.1 F4.5
N6 X-1.5
N7 Y-1.5
N8 G80 G00 Z1. M09
N9 G53 G49 Z0. M05
N10 M30
%
```

Manufacturing CNC's Night of Magic

Haas CNC Mill
G87 Bore In, Manual Retract Canned Cycle

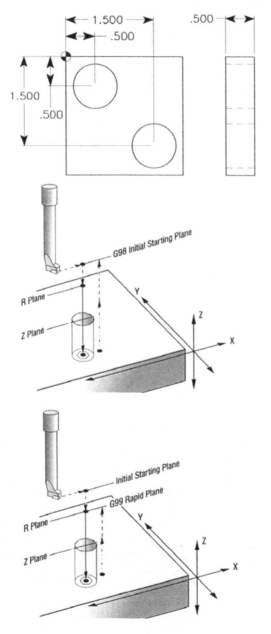

Haas CNC Mill
G88 Bore In, Dwell, Manual Retract Canned Cycle

X   Rapid X-axis location. (optional)
Y   Rapid Y-axis location. (optional)
Z   Z-depth. (Feed to Z-depth starting from R-plane)
R   R-plane. (Rapid point to start feeding)
F   Feed-Rate in inches (mm) per minute.

This G code is modal so that it is activated every X and/or Y-axis move, and it will rapid to that position and then cause this canned cycle to be executed again, until it's canceled. Use G98 and G99 for the Z position clearance location between holes.

```
%
o10087 (G88 Bore in, dwell, manual retract cycle)
N1 T19 M06 (Boring bar)
N2 G90 G54 G00 X0.00 X1.5 Y1.5
N3 S1450 M03
N4 G43 H19 Z1. M08
N5 G88 G99 Z-.42 P0.2 R0.1 F4.5
N6 X-0.5 Y-1.5
N7 G80 G00 Z1. M09
N8 G53 G49 Z0. M05
N9 M30
%
```

Manufacturing CNC's Night of Magic

Haas CNC Mill
G88 Bore In, Dwell, Manual Retract Canned Cycle

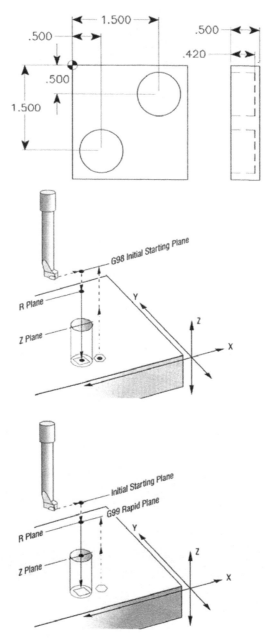

Haas CNC Mill
G89 Bore In, Dwell, Bore Out Canned Cycle

X   Rapid X-axis location. (optional)
Y   Rapid Y-axis location. (optional)
Z   Z-depth. (Feed to Z-depth starting from R-plane)
R   R-plane. (Rapid point to start feeding)
F   Feed-Rate in inches (mm) per minute.

This G code is modal so that it is activated every X and/or Y-axis move, and it will rapid to that position and then cause this canned cycle to be executed again, until it's canceled. A dwell in this cycle in seconds"," milliseconds will happen at the end of the Z-depth with P defined. Use G98 and G99 for the Z position clearance location between holes.

```
%
o10088 (G89 Bore in, dwell, bore out cycle)
N1 T19 M06 (Boring bar)
N2 G90 G54 G00 X1.625 Y-0.375
N3 S1450 M03
N4 G43 H19 Z1. M08
N5 G89 G99 Z-0.375 P0.2 R0.1 F4.5
N6 X1. Y-1.
N7 X0.375 Y-1.625
N8 G80 G00 Z1. M09
N9 G53 G49 Z0. M05
N10 M30
%
```

Mastercam Class

Haas CNC Mill
G89 Bore In, Dwell, Bore Out Canned Cycle

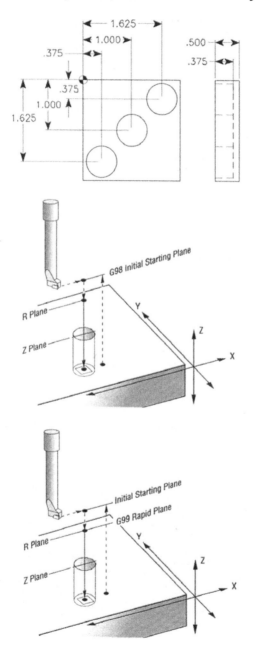

Haas CNC Mill
G90 Absolute & G91 Incremental Position Command

G90 Absolute Position Command:

In absolute positioning, all coordinate positions are given with regard to their relationship to a fixed zero, origin point, that is referred to as part zero. This is the most common type of positioning.

When using a G90 absolute position command, each dimension or move is referenced from a fixed point, known as Absolute Zero (part zero). Absolute Zero is usually set at the corner edge of a part, or at the center of a square or round part, or an existing bore. Absolute Zero is where the dimensions of a part program are defined from.

Absolute dimensions are referenced from a known point on the part, and can be any point the operator chooses, such as the upper-left corner, center of a round part, or an existing bore.

The Key to understanding Absolute dimensions is that they are always in reference to the Absolute Zero (part zero). This part zero (work offset G codes G54 – G59 and G110 – G129) are set by the operator in the offset display using the Handle Jog operation mode. It can also be switched to a new part zero position during the program using a different work offset G code that defines in it, another location (when machining with multiple vises and / or fixtures at separate locations on the machine table).

Each dimension, or X – Y point is known as a coordinate. If a position 2-inches to the right, and 2-inches down (toward you) from part zero was programmed, the X coordinate would be X2.0 and the Y coordinate would be Y-2.0. And the machine would go to that exact location from part zero, regardless of where it began, within the travel of the machine tool. X2.0 Y-2.0 could be a hole location, an arc end point, or the end of a line which are known coordinate vales.

G91 Incremental Position Command:

Incremental positioning concerns itself with distance and direction from the last position. A new coordinate is entered in terms of its relationship to the previous position, and not from a fixed zero or origin. In other words, after a block of information has been executed, the position that the tool is now at is the new zero point for the next move to be made.

This code is model and changes the way axis motion commands are interpreted. G91 makes all subsequent commands incremental.

Incremental dimensions are referenced from one point to another. This can be a convenient way to input dimensions into a program (especially from G81 – G89, G73, G74 and G77 canned cycles depending on the blueprint.

When using a G91 incremental position command. Each measurement or move is the actual distance to the next location (whether it is a hole location, end of arc, or end of line) and is always in reference from the current location.

If you programmed a G91 with an X coordinate of X2.0 and a Y coordinate of Y-2.0, the machine would go that exact distance from where it is, regardless of where it began, within the travel of the machine tool.

Absolute mode should be your positioning mode of choice for most applications. There are times when incremental mode can be quite helpful. Repeating motions within a subroutine, foe example, is one excellent example. If you have six identical pockets to machine on a Haas mill, you can save programming effort if you specify the motions incrementally to machine one pocket. Then just call up the subroutine again to repeat the commands to do another pocket at a new location.

Haas CNC Mill
G101 Enable Mirror Image

Programmable mirror image can be turned on or off individually for any of the axes. When one is ON, axis motion may be mirrored, or reversed, around the work zero point. These G codes should be used in a command block without any other G codes. They do not cause any axis motion. The bottom of the screen will indicate when an axis is mirrored. Also see Settings 45 through 48 for mirror imaging.

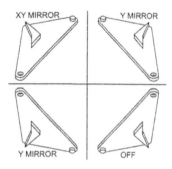

The format for turning mirror image on or off is.

G101 (will turn ON mirror imaging for the X-axis)

G100 (will turn OFF mirror imaging for the X-axis)

Most mirror image applications would consist of irregular pockets and contours and would most likely be set up in subprograms for convenience.

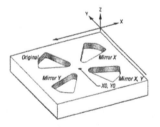

Note: after completion of the first item, a Z-axis clearance move should be made. Then, the mirror image should be turned on with an axis specification. The following line needs the coordinates of the starting location of the original pocket. The following line will feed to the required Z-axis depth, the next line would contain a subprogram call or a contour definition, and last, a positive Z-axis clearance move.

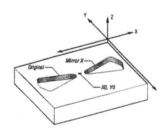

Note: when milling a shape with X-Y motions, turning ON mirror image for just one of the X and Y will change climb milling to conventional milling and/or conventional milling to climb milling. As a result, you may not get the type of cut or finish that was desired. Mirror image of both X and Y will eliminate this problem

Haas CNC Mill
G101 Enable Mirror Image

Mirror Image and Cutter Compensation:

When using cutter compensation with mirror imaging, follow this guideline. After turning mirror imaging on or off with G100 or G101, the next motion block should be to a different work coordinate position than the first one.

The following code is an example.

Correct:
G41 X1.0 Y1.0
G01 X2.0 Y2.0
G101 X0
G00 Z1.0
G00 X1.0
G00 X2.0 Y2.0
G40

Incorrect:
G41 X1.0 Y1.0
G01 X2.0 Y2.0
G101 X0
G00 Z1.0
G00 X2.0 Y2.0
G40

Note: mirroring only one of the X or Y axis will cause the cutter to move along the opposite side of a cut. In addition, if mirror imaging is selected for only one axis of a circular motion plane G02, G03 then they are reversed, and left and right cutter compensation commands G41, G42 are reversed.

Note: when milling a shape with XY motions, turning ON mirror image for just one X or Y-axis will change climb milling to conventional milling and / or conventional milling to climb milling. As a result, you may not get the type of cut or finish that was desired. Mirror imaging of both X and Y will eliminate this problem.

The Pockets should be arranged around a given origin.
Usually described as X0, Y0.

Materials & Process Class

Haas CNC Mill
G101 Enable Mirror Image

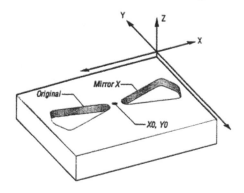

Image: program o3600 and subprogram o3601

```
%
o3600 (mirror image X-axis)
T1 M06 (1/4" Dia. end-mill)
G00 G90 G54 X-.4653 Y.052 S5000 M03
G43 H01 Z.1 M08
G01 Z-.25 F5.
F20.
M98 P3601
G00 Z.1
G101 X0.
X-.4653 Y.052
G01 Z-.25 F5.
F20.0
M98 P3601
G00 Z.1
G100 X0.
G28 G91 Y0 Z0
M30
%
```

```
%
o3601 (contour subprogram)
G01 X-1.2153 Y.552
G03 X-1.3059 Y.528 R.0625
G01 X-1.5559 Y.028
G03 X-1.5559 Y-.028 R.0625
G01 X-1.3059 Y-.528
G03 X-1.2153 Y-.552 R.0625
G01 X-.4653 Y-.052
G03 X-.4653 Y.052 R.0625
M99
%
```

Haas CNC Mill
G101 Enable Mirror Image

Image: program o3700 and subprogram o3701

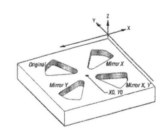

%
o3700 (mirror image X, Y, and XY axes)
T1 M06 (1/4" Dia. end-mill)
G00 G90 G54 X-.2923 Y.3658 S5000 M03
G43 H01 Z.1 M08
G01 Z-.25 F5.
F20.
M98 P3701 (pocket contour subprogram call)
G00 Z.1 (part clearance)
G101 X0. (turn on mirror image X-axis)
X-.1923 Y.3658 (position to original coordinates)
G01 Z-.25 F5. (feed to Z-depth)
F20. (pocket feed-rate)
M98 P3701 (pocket contour subprogram call)
G00 Z.1
G100 X0. (cancel mirror image X-axis)
G101 Y0. (turn on mirror image Y-axis)
X-.2923 Y.3658
G01 Z-.25 F5.
F20.
M98 P3701 (pocket contour subprogram call)
G00 Z.1
G100 Y0. (cancel mirror image Y-axis)
G101 X0. Y0. (turn on mirror image X, Y, axis)
X-2923 Y.3658
G01 Z-.25 F5.
F20.
M98 P3701
(pocket contour subprogram call)
G00 Z.1
G100 X0. Y0.
G28 G91 Y0. Z0.
M30
%

%
o3701 (contour subprogram)
G01 X-.469 Y1.2497
G03 X-.5501 Y1.2967 R.0625
G01 X-1.0804 Y1.12
G03 X-1.12 Y1.0804 R0.0625
G01 X-1.2967 Y.5501
G03 X-1.2497 Y.469 R.0625
G01 X-.3658 Y.2923
G03 X-.2923 Y.3658 R.0625
M99
%

Haas CNC Mill
G150 General Purpose Pocket Milling

P   Subprogram call of program number that defines pocket geometry.
X   X position of starting hole. (not needed if you're already at that location)
Y   Y position of starting hole. (not needed if you're already at that location)
Z   Final depth of pocket.
Q   Incremental Z-axis depth, starting from R plane.
R   R plane clearance position.
I   X-axis shift over cut increment (use I or J, not both).
J   Y-axis shift over cut increment (use I or J, not both).
K   Finishing cut allowance.
G41 Left of program path, cutter compensation.
G42 Right of program path, cutter compensation.
D   Cutter-comp geometry offset register number.
F   Feed-rate

This G code provides for general purpose pocket milling. The shape of the pocket to be machined must be defined by a series of motions within a subprogram. A series of motions in either the X or Y-axis will be used to cut out the specified shape, followed by a finishing pass to clean up the outer edge. Only one of either I or J must be specified. If I is used, the pocket is cut by shifting over in the X-axis, and the strokes will be along the y-axis. If J is used, shifting over in the Y-axis cuts the pocket, and the strokes will be along the X-axis. I and J must be positive numbers. The K command is a finish pass and must be a positive number. There is no finishing pass for the Z depth using the K command. A Z-depth finishing pass can be calculated using Z, Q, and R, commands.

Multiple Q cuts can be defined to get to the final Z-depth. The feed down to the final Z-depth starts from the R-plane, feeding down by the Q amount for each pass until the Z-depth is reached. Q must be positive.

If an L count is specified on the G150 command line, with a G91 and an X and/or Y value, the entire block is repeated again at an incremental X or Y distance, L number of times.

The subroutine must define a closed loop are with a series of G01, G02, or G03 motions in the X and Y-axis, and must end with an M99. G codes G90 and G91 can also be used in the subprogram to select absolute or incremental motion. Any codes other than G, I, J, R, X, or Y, are ignored in the subprogram, which must consist of no more than 40 moves.

In control software version 11.11, G150 pocket milling was increased to 40 geometry moves. In all previous versions, G150 could be defined with no more than 20 geometry moves.

With a G150 pocket milling command, you may need a clearance hole drilled to Z-depth of pocket, prior to the end-mill entering for a G150 pocket cycle, since it plunges straight down in the Z-axis. You may choose to specify an XY starting hole location to be drilled, prior to, or in the G150 command

Haas CNC Mill
G150 General Purpose Pocket Milling

The first motion of a subprogram should define a move from the starting hole location onto a start point of the pocket geometry. The final move in the subprogram should close the loop at the same point where you began the pocket geometry. In the example on the following page, the start point of the G150 is X1.5 Y3.25, and the first move of the subprogram is X0.5.

Therefore, the last move of the subprogram must return to X0.5 Y3.25.

If K is specified, the roughing cuts will cut inside the programmed pocket size by the amount of K. the finishing pass will follow along the pocket geometry edge and is done at the full pocket depth.

```
%
o10110 (program name hear)
N101 T16 M06 (1/2" Dia. Drill)
N102 G90 G54 G00 X1.5 Y3.25 (drilling start location)
N103 S1528 M03
N104 G43 H16 Z1. M08
N105 G83 G99 Z-1.25 Q0.25 R0.1 F6.1
N106 G80 G00 Z1. M09
N107 G53 G49 Z0. M05
N108 T11 M06
N109 G90 G54 G00 X1.5 Y3.25
(start location for pocket milling routine)
N110 S2100 M03
N111 G43 H11 Z1. M08
N112 G01 Z.01 F30. (feed down to start point in Z)
N113 G150 P10111 G41 D11 I.4 (or J.4) K.02 Z-1.25 Q.63 R.01 F16.5
N114 G40 G01 X1.5 Y3.25 (cancel cutter-comp)
N115 G00 Z0.1 M09
N116 G53 G49 Y0. Z0. M05
N117 M30
%
```

The G150 pocket-milling sub-program is called up on line N113 with G150 P1011. This tells the control to look for program o10111 for the finish geometry of the pocket being machined.

(Continued on next page)

Haas CNC Mill
G150 General Purpose Pocket Milling
(Continued from previous page)

%
o10111
N1 G01 X0.5
N2 Y1.5
N3 G03 X2.25 R0.875
N4 G01 X5.25
N5 G03 X7. R0.875
N6 G01 Y5.
N7 G03 X5.25 R0.875
N8 G01 X2.25
N9 G03 X0.5 R0.875
N10 G01 Y3.25
N11 M99
%

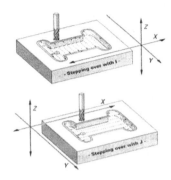

%
o10112 (program name hear)
N101 T16 M06 (1/2" Dia. Drill)
N102 G90 G54 G00 X1.5 Y3.25
(drilling start location)
N103 S1528 M03
N104 G43 H16 Z1. M08
N105 G83 G99 Z-1.25 Q0.25 R0.1 F6.1
N106 G80 G00 Z1. M09
N107 G53 G49 Z0. M05
N108 T11 M06
N109 G90 G54 G00 X1.5 Y3.25
(start location for pocket milling routine)
N110 S2100 M03
N111 G43 H11 Z0.1 M08
N112 G01 Z0.01 F30.
 (feed down to start point in Z)
N113 G150 P10113 G41 D11 I.4 (or J.4) .
. . . K.02 Z-1.25 Q.63 R.01 F16.5
N114 G40 G01 X1.5 Y3.25
(cancel cutter-comp)
N115 G00 Z0.1 M09
N116 G53 G49 Y0. Z0. M05
N117 M30
%

The G150 pocket-milling sub-program is called up on line N113 with G150 P1013. This tells the control to look for program o10111 for the finish geometry of the pocket being machined.

%
o10113
N1 G01 X0.5
N2 Y1.5
N3 G03 X2.25 R0.875
N4 G01 X4.
N5 Y3.
N6 X2.5
N7 G02 Y3.5 R0.25
N8 G01 X5.
N9 G02 Y3. R0.25
N10 G01 x3.4
N11 Y1.5
N12 X5.25
N13 G03 X7. R0.875
N14 G01 Y5.
N15 G03 X5.25 R0.875
N16 G01 X2.25
N17 G03 X0.5 R0.875
N18 G01 Y3.25
N19 M99 (returns to main program)
%

Haas CNC Mill
G150 Pocket Milling Notes:

(Continued from previous page)

1) Position the end-mill to the starting point inside the pocket. Do not have cutter overlap on any line of the pocket geometry with the entry of the end-mill.

2) The g150 command line calls up a subprogram with a P command (P12345) that specifies the separate subprogram (o12345) which defines the geometry of a pocket. This pocket geometry must be defined in 40 strokes (moves) or less. In control software version 11.11, G150 pocket milling was increased to 40 geometry moves. In all previous versions, a G150 could not be defined with more than 20 geometry moves in the subprogram called up by the G150.

3) The G150 pocket geometry command cannot be listed after the M30 command, like the way M97 (local subprogram call) is used. Instead, G150 uses a P command to call up a separate program, which contains the pocket geometry.

4) You can also define an island within the G150 pocket command. Remember that you only have a total of 40 moves to define the pocket geometry (only 20 moves for machines with version 11.11 and earlier).

5) When positioning on and off of an island inside a G150 subprogram that's using cutter compensation (G41 / G42). The cutter will be either left or right of the line you're positioning onto. If you're positioning onto the island or pocket at the same location you position off the island or pocket, you will leave a large scallop the size of the offset radius. To eliminate the scallop, you need to overlap the entry and exit moves by a little more than the offset amount, in the offset register, so that the center of the cutter overlaps the path on the entry and exit moves, when defining an island in a G150 pocket command.

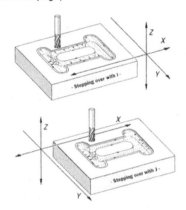

6) Define the pocket geometry tool path in the direction you wish to cut. Define the path counterclockwise to climb cut using G41, or clockwise to conventional cut with a G42. The cutting path direction needs to be reversed when positioning onto an island inside the pocket when climb or conventional cutting with end-mill.

7) When you're defining the pocket geometry in the subprogram, define it from a starting point inside the pocket. Then define the pocket geometry, beginning with a move onto the side of the pocket. Close the loop where you began, and end the subprogram there with an M99. Don't define a move back to the starting point where you began.

8) The Q value must be defined in a G150 command, even if you're only doing one pass to the final Z-depth. The Q amount is defined from the R-plane. If you only want one pass, and your final depth is X-0.25, and you're starting the pocket at R0.1 above part, then your Q value will be Q0.35. If, however, you give it a Q0.25, it will move from R0.1 down to Z-0.15 (the amount Q0.25) into the pocket. Then it will take a second pass to the final depth of Z-0.25.

Haas CNC Mill
G150 Pocket Milling Notes:

9) If you execute a G150 pocket milling command and have an invalid move in your pocket geometry with an alarm 370, (pocket definition error), the control will stop on the G150 command line in the main program. To locate this invalid move quicker (within the pocket geometry). Run the pocket subprogram separately in Graphics to find the errors.

To run the subprogram in Graphics, you'll need to enter a feed-rate (G01) command on the first line of the subprogram. Troubleshoot the pocket geometry in Graphics.

After correcting the moves in the subprogram, take the feed-rate command back out, and verify the fix in Graphics on the main program

G150 General Purpose Pocket Milling

```
%
o01001
T1 M6
G90 G54 G00 X0. Y1.5
S2000 M03
G43 H01 Z0.1 M08
G01 Z0.01 F30.
G150 P1002 Z-0.5 Q0.25 R0.01 J0.3 . . .
. . . K0.01 G41 D01 F10.
G40 G01 X0. Y1.5
G00 Z1.0 M09
G53 G49 Y0. Z0.
M30
%
```

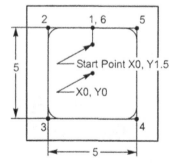

(Continued on next page)

Haas CNC Mill
G150 General Purpose Pocket Milling
(Continued from previous page)

Subprogram
%
o01002
G01 Y2. (1)
X-2.5 (2)
Y-2.5 (3)
X2.5 (4)
Y2.5 (5)
X0. (6) (close pocket loop)
M99 (return to main program)
%

Absolute Subprogram
%
o01002
G90 G01 Y2. (1)
X-2.5 (2)
Y-2.5 (3)
X2.5 (4)
Y2.5 (5)
X0. (6)
M99 (return to main program)
%

Incremental Subprogram
%
o01002
G91 G01 Y0.5 (1)
X-2.5 (2)
Y-5. (3)
X5. (4)
Y5. (5)
X-2.5 (6)
G90
M99 (return to main program)
%

Haas CNC Mill
G150 General Purpose Pocket Milling

%
o02010
T1 M6 (1/2" Dia. end-mill)
G90 G54 G00 X2. Y2. (XY start point)
S2500 M03
G43 H01 Z0.1 M08
G01 Z0.01 F30.
G150 P2020 X2. Y2. Z-0.5 Q0.5 R0.01 I0.3 K0.01 G41 D01 F10.
G40 G01 X2. Y2.
G00 Z1.0 M09
G53 G49 Y0. Z0.
M30
%

Subprogram
%
o02020
G01 Y1. (1)
X6. (2)
Y6. (3)
X1. (4)
Y3.2 (5)
X2.75 (6)
Y4.25 (7)
X4.25 (8)
Y2.75 (9)
X2.75 (10)
Y3.8 (11)
X1. (12)
Y1. (13)
X2. (14) (close pocket loop)
M99 (return to main program)
%

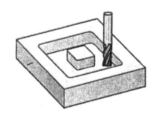

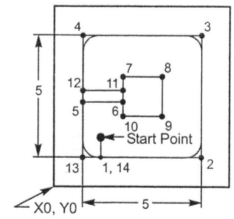

Haas CNC Mill
G150 General Purpose Pocket Milling

%
o03010
T1 M6 (1/2" Dia. end-mill)
G90 G54 G00 X2. Y2. (XY start point)
S2500 M03
G43 H01 Z0.1 M08
G01 Z0. F30.
G150 P3020 X2. Y2. Z-0.5 Q0.5 R0.01 J0.3 K0.01 G41 D01 F10.
G40 G01 X2. Y2.
G00 Z1.0 M09
G53 G49 Y0. Z0.
M30
%

Subprogram
%
o03020
G01 Y1. (1)
X6. (2)
Y6. (3)
X1. (4)
Y3.5 (5)
X2.5 (6)
G02 I1. (7)
G02 X3.5 Y4.5 R1. (8)
G01 Y6. (9)
X1. (10)
Y1. (11)
X2. (12) (close pocket loop)
M99 (return to main program)
%

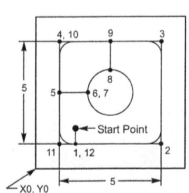

Haas CNC Mill
G187 Accuracy Control of Corners

The G87 code is used to select the accuracy with which corners are machined. The format for using G187 is G187 Ennnn, where nnnn is the desired accuracy. Refer to "Contouring Accuracy" for more information.
G187 E0.01 (sets G187 to set value)
G187 (reverts to Setting 85 value)

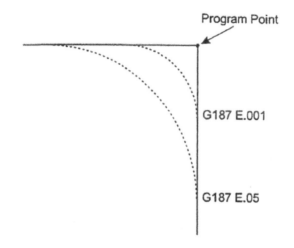

Mastercam Lab

Haas CNC Mill
Safety Procedures

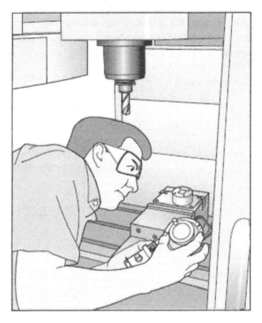

Professional Machine Operators
Always Consider
Safety First

Haas CNC Mill
Safety Procedures
Read Before Operating This Machine

Only authorized personnel should work on this machine. Untrained personnel present a hazard to themselves and the machine and improper operation will void the warranty.

Check for damaged parts and tools before operating the machine. Any part or tool that is damaged should be properly repaired or replaced by authorized personnel. Do not operate the machine if any component does not appear to be functioning correctly. Contact your shop supervisor.

Use appropriate eye and ear protection while operating the machine. ANSI-approved impact safety goggles and OSHA-approved ear protection is recommended to reduce the risks of sight damage and hearing loss.

Do not operate the machine unless the doors are closed and the door interlocks are functioning properly. Rotating cutting tools can cause severe injury. When a program is running, the mill table and spindle head can move rapidly at any time in any direction.

The Emergency Stop button (also known as an Emergency Power Off button) is the large. circular red switch located on the Control Panel. Pressing the Emergency Stop button will instantly stop all motion of the machine. the servo motors. the tool changer. and the coolant pump. Use the Emergency Stop button only in emergencies to avoid crashing the machine.

The electrical panel should be closed and the key and latches on the control cabinet should be secured at all times except during installation and service . At those times. only qualified electricians should have access to the panel. When the main circuit breaker is on. There is high voltage throughout the electrical panel (including the circuit boards and logic circuits) and some components operate at high temperatures.

Therefore extreme caution is required. Once the machine is installed the control cabinet must be locked and the key available only to qualified service personnel.

Consult your local safety codes and regulations before operating the machine. Contact you dealer anytime safety issues need to be addressed.

DO NOT modify or alter this equipment in any way. If modifications are necessary, all such requests must be handled by Haas Automation, Inc. Any modification or alteration of any Haas Milling or Turning Center could lead to personal injury and/or mechanical damage and will void your warranty.

It is the shop owner's responsibility to make sure that everyone who is involved in installing and operating the machine is thoroughly acquainted with the installation operation and safety instructions provided with the machine BEFORE they perform any actual work. The ultimate responsibility for safety rests with the shop owner and the individuals who work with the machine.

Haas CNC Mill
Safety Procedures
Observe All Of The Warnings And Cautions Below

This machine is automatically controlled and may start at any time.

This machine can cause severe bodily injury.

Do not operate with the doors open.

Avoid entering the machine enclosure.

Do not operate without proper training.

Always wear safety goggles.

Never place your hand on the tool In the spindle and press ATC FWD, ATC REV, NEXT TOOL, or cause a tool change cycle. The tool changer will move In and crush your hand.

To avoid tool changer damage, ensure that tools are properly aligned with the spindle drive lugs when loading tools.

The electrical power must meet the specifications In this manual. Attempting to run the machine from any other source can cause severe damage and will void the warranty.

DO NOT press POWER UP/RESTART on the control panel until after the Installation Is complete.

DO NOT attempt to operate the machine before all of the Installation Instructions have been completed.

NEVER service the machine with the power connected.

Improperly clamped parts machined at high speeds/feeds may be ejected and puncture the safety door. Machining oversized or marginally clamped parts Is not safe.

Windows must be replaced If damaged or severely scratched. Replace damaged Windows Immediately.

Do not process toxic or flammable material. Deadly fumes can be present. Consult material manufacturer for safe handling of material by-products before processing.

The spindle head can drop without notice. Personnel must avoid the area directly under the spindle head.

Unattended Operation

Fully enclosed Haas CNC machines are designed to operate unattended: however, your machining process may not be safe to operate unmonitored.

As it is the shop owner's responsibility to set up the machines safely and use best practice machining techniques it is also their responsibility to manage the progress of these methods. The machining process must be monitored to prevent damage if a hazardous condition occurs.

For example, if there is the risk of fire due to the material machined then an appropriate fire suppression system must be installed to reduce the risk of harm to personnel, equipment and the building. A suitable specialist must be contacted to install monitoring tools before machines are allowed to run unattended.

It is especially important to select monitoring equipment that can immediately perform an appropriate action without human intervention to prevent an accident should a problem be detected.

Haas CNC Mill
Milling Machine Travels

20 – 30 Inch VMCS

VF-0		Gantry	
VF-1	20" x 16" x 20"	GR-408	50" x 100" x 11"
VF-2	30" x 16" x 20"	GR-510	121" x 61" x 11"
VF-1YT	20" x 20" x 20"	GR-512	145" x 61" x 11"
VF-2YT	30" x 20" x 20"	GR710	121" x 85" x 11"
		GR-712	145" x 85" x 11"

40 – 64 Inch VMCS

		High-Speed	
VF-3	40" x 20" x 25"	VF-2SS	30" x 16" x 20"
VF-3YT	40" x 26" x 25"	VF2SSYT	30" x 20" x 20"
VF-4	50" x 20" x 25"	VF-3SS	40" x 20" x 20"
VF-5/40	50" x 26" x 25"	VF-3SSYT	40" x 26" x 25"
VF-5/40XT	60" x 26" x 25"	VF-4SS	50" x 20" x 25"
VF-6/40	64" x 32" x 30"	VF-6SS	64" x 32" x 30"
VF-8/40	64" x 40" x 30"		

80 – 120 Inch VMCS

		Pallet Changer Machines	
VF-7/40	84" x 32" x 30"	VF-4SSAPC	50" x 20" x 25"
VF-9/40	84" x 40" x 30"	VF-3APC	40" x 20" x 25"
VF-10/40	120" x 32" x 30"	VF-3SSAPC	40" x 20" x 25"
VF-11/40	120" x 40" x 30"	VF-4APC	50" x 20" x 25"

Mold Machine

		Mini-FMS (Dual Pallet Changers)	
VM-2	30" x 20" x 20"	VF-4SSAPCQ	50" x 20" x 25"
VM-3	40" x 26" x 25"	VF3APCQ	40" x 20" x 25"
VM-6	64" x 32" x 30"	VF3SSAPCQ	40" x 20" x 25"
		VF-4APCQ	50" x 20" x 25"

Mill Drill Center

		5-Axis Trunnion	
MDC-500	20" x 14" x 20"	VF-5/40TR	38" x 26" x 25"
		VF-2TR	30" x 16" x 20"
		VF-6/40 TR	64" x 32" x 30"

Tool-Room Mills

		5-Axis Spindle	
TM-1P	30" x 12" x 16"	VR-8	64" x 40" x 30"
TM-1	30" x 12" x 16"	VR-9	84" x 40" x 30"
TM-2	40" x 16" x 16"	VR-11B	120" x 40" x 30"

Mini VMCS

Mini-Mill	16" x 12" x 10"
Super Mini Mill	16" x 12" x 10"

Haas CNC Mill
Milling Machine Travels

Medium Frame / 50-Taper
VF-3YT/50	40" x 26" x 25"
VF-5/50	50" x 26" x 25"
VF-50/50XT	60" x 26" x 25"
VF-6/50	64" x 32" x 30"
VF-8/50	64" x 40" x 30"

Large Frame / 50-Taper
VF-7/50	84" x 32" x 30"
VF-9/50	84" x 40" x 30"
VF-10/50	120" x 32" x 30"
VF-11/50	120" x 40" x 30"

X-Large Frame
VS-1	84" x 50" x 50"
VS-3	150" x 50" x 50"

50-Taper 5-Axis / 50-Taper
VF-5/50TR	38" x 26" x 25"
VF-6/50TR	64" x 32" x 30"

Office Mill
OM-1	8" x 8" x 8"
OM-1A	8" x 8" x 12"
OM-2	12" x 10" x 12"
OM-2A	12" x 10" x 12"

HMC Horizontal Spindle EC-Series
EC-300	20" x 18" x 14"
EC-400	20" x 20" x 20"
EC-400PP	20" x 20" x 20"
EC-500	32" x 20" x 28"
EC-1600	64" x 40" x 32"
EC-2000	84" x 40" x 32"
EC-3000	120" x 40" x 32"

HMC Horizontal Spindle HS-Series
HS-3	150" x 50" x 60"
HS-3R	150" x 50" x 60"
HS-4	150" x 66" x 60"
HS-4R	150" x 66" x 60"
HS-6	84" x 50" x 60"
HS-6R	84" x 50" x 60"
HS-7	84" x 66" x 60"
HS-7R	84" x 66" x 60"

Haas CNC Mill
Quick Start Up Guide

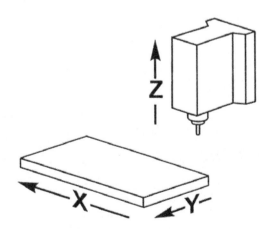

Haas CNC Mill: Home Position

Haas CNC Mill
Quick Start Up Guide
Start Up & Zeroing – Out Machine

The steps outlined below are for starting up the Haas VF-0 or Mini-Mill CNC Machining Centers. These steps serve as an overall guide to the beginning operator. Though intended to be used for Haas CNC machining centers you will find in industry.

Note: DO NOT attempt the following procedures until they, are first demonstrated by the Instructor.

1) Check the air pressure being delivered to the machine, it should read 85-90 PSI. The regulator is located at the rear of machine.

2) Rotate the main breaker at the rear of the machine to the ON position.

3) Clean the spindle opening. Wipe spindle taper clean with a clean cloth rag and apply light oil. (See instructor for oil)

4) Rotate clockwise and release the red EMERGENCY-STOP Switch. This allows power to the axes servo motors, tool changer, spindle drive, and coolant pump.
Location: control Panel

5) Press the green POWER ON button.
Location: Control Panel

6) Clear the flashing ALARM message by pressing the red RESET button. The screen should then display a NO ALARM message. *Location: Control Panel Keypad*

7) Press the POSIT (position) button to view the machine position screen. *Location: Control Panel Keypad*
Note: Make Sure the doors are in the closed position
before continuing.

8) To zero-out the machine to its home position, start by pressing the ZERO RET (zero-return) button. This puts the machine in zero-return mode. *Location: Control Panel*

9) Press the AUTO ALL AXES button. This will initiate the zero-return sequence for the X, Y and Z axes.
Location: Control Panel Keypad

10) Confirm that all axes are at machine zero and the controller has reset it's machine zero location by viewing the machine position screen. *Location: Control Panel Screen*

CMM
Coordinate Measuring Machine

Haas CNC Mill
Quick Start Up Guide
Jog Commands & Aligning a Vise

The steps outlined below are for aligning a vise on the Haas VF-0 or Mini Mill CNC Machining Center. Though intended to be used on a Haas, these concepts may be applied to other CNC machining centers you will find in industry.

Note: DO NOT attempt the following procedures until they, are first demonstrated by the Instructor.

1) Start up and zero out (home) the machine (If it is not already zeroed out).

2) Clean any chips or debris off the vise flange and machine table. Loosen the vise and move it out of alignment. Show the instructor the vise before continuing.

3) Lightly tighten one side of the vise. Tighten the other side finger tight.

4) Assemble a dial test indicator and magnetic base tool holder.

5) Mount magnetic base to the tool head.

6) On the KEYPAD press the "Hand Jog" key.

7) On the KEYPAD press the "X, Y or Z" axis key. It does not matter if you select + or – at this time since the JOG HANDLE rotation will determine which direction the axis will move.

8) On the KEYPAD select the jog handle step increment by pressing the ".0001", ".001", ".01", or ".1" key.

9) Jog the X and Y axes to the desired general location you intend to start from. Location should place the indicator in front of the vise jaw and towards one end.

Note: As a General Rule: 0.0001" is used for precise locating of the axis and for extremely slow, controlled movement.

Note: As a General Rule: 0.001" is used for semi-precise locating of the axis and for slow, controlled movement.

Note: As a General Rule: 0.01" is used for general locating of the axis. Some caution is required around machine travel limits and part, fixture or other obstructions.

Note: As a General Rule: 0.1" is used for rapid positioning of the axis. Caution is required around machine travel limits and part, fixture or other obstructions. Moving in a slight delay of the desired machine response.

10) Now move the Z-axis towards the vise.

11) Put the jog control into .001" step increment and slowly position the indicator to the back-jaw of the vise, on the same side, which is tightened.

12) Begin to align the vise. Be sure the control is set to move the X axis before jogging. Jog the table in the X axis right and left and adjust vise until there is .001 maximum run-out.

13) Tighten the vise hold down nuts. Have the instructor check the vise alignment.

14) Remove the magnetic base and indicator.

15) Return the Z axis to machine zero.

Press: ZERO RET (Zero Return).

Enter: Z to select axis. (Press cancel if you push the wrong button).

Press: ZERO SINGL AXIS (Zero Signal Axis), The Z-axis should return to it's machine zero location.

16) Shut down the machine. Press: POWER OFF & EMERGENCY STOP

Haas CNC Mill
Quick Start Up Guide
Finding & Storing Work Offsets

The steps outlined below are for storing work offsets on a Haas VF-0 or Mini Mill Machining Centers. The same concepts may be applied to other CNC machining centers found in industry.

Note: DO NOT attempt the following procedures until they, are first demonstrated by the Instructor.

1) Start up and zero out (home) the machine (if it is not already zeroed out).

1.1) Secure the part on the machine.

2) Secure the edge finding tool in a holder or drill chuck and install in the machine spindle.

2.1) Hold the tool holder with your left hand.

2.2) Press: TOOL RELEASE button located on the controller and hold. You should hear a hissing sound coming from the spindle. If not, make sure the machine is in the jog mode.

2.3) Place the tool holder in the spindle and release the TOOL RELEASE button. Make sure it is aligned and installed correctly.

3) Set The RPM

3.1) Press: MDI (Manual Data Input).

3.2) Press: PROGRAM / CONVRS (Program).

3.3) Press: DELETE (if needed to clear old commands).

3.4) Enter: S1000 M03, if you push wrong key? Press: CANCEL.

3.5) Press: EOB (End of Block).

3.6) Press: WRITE / ENTER.

3.7) Close the machine doors.

3.8) Press: CYCLE START. The spindle should now be rotating clockwise at 1000 RPM.

4) Jog the Z-axis down towards the part.

4.1) Press: HAND JOG.

4.2) Select the .010" jog increment by pressing: .01

4.3) Jog X and Y to location near the near the part.

4.4) Lower the Z-axis to approximately 1.0" above the part, by rotating the jog handle CCW in the minus (-) direction.

5) Select the offset register display.

5.1) Press: OFSET (Offset).

5.2) Press: PAGE UP or PAGE DOWN until you see G54 in the WORK-ZERO OFFSET register.

5.3) If necessary, move the cursor to G54.

6) Pick-up the X axis part edge with edge finder.

6.1) Jog the edge finding tool into position with the Z axis about .20 inch below the top surface of the part.

6.2) Using the Jog Handle and in .001" inch step increment, pick-up the X axis edge with the edge finder.

6.3) After locating the edge: Raise the edge finding tool about .5 inch above the part. Jog until the edge is centered on the spindle axis.

Note: The distance required equals one half the edge finding tool diameter. Example: Press .1 and move handle one graduation −Y for a .20 diameter edge-finder.

Haas CNC Mill
Quick Start Up Guide
Finding & Storing Work Offsets

7) Store the work zero offset. Press: PART SERO SET to store the X-axis value in the work offset register. Check to make sure the X-axis position (bottom of screen) matches the value stored in the G54 X-axis register.

8) Pick-up the Y-axis part edge.

8.1) Jog the edge-finding tool into position with the Z-axis about .200 inches below the top surface of the part.

8.2) Using the Jog-Handle and in .001" step increment, pick-up the Y-axis edge.

8.3) After locating the edge with the edge-finder raise the tool about 5 inches above the part and jog until the edge is centered on the spindle-axis

Note: The distance required equals one half the edge finding tool Dia.

9) Store the work offset: Press PART ZERO SET to store the Y-axis value in the work offset register. Check to make sure the Y-axis position (bottom of screen) matches the value stored in the G54 Y-axis register.

10) Check the WORK ZERO OFFSET location.

10.1) Move: The Z-axis to .500" above the part.

10.2) Press: MDI (Manual Data Input) If necessary, clear the screen of any commands displayed by pressing DELETE.

10.3) Enter: G90 G54 X0 Y0

10.4) Press: EOB (End of Block)

10.5) Press: WRITE

10.6) Close: The machine doors.

10.7) Press: CYCLE START. The table should move to the work coordinate position stored in the G54 register.

10.8) Check: The location by comparing the spindle-axis and part zero location. They should be in alignment if not, repeat steps 6 – 10 again.

11) Raise the spindle to machine zero.

11.1) Press: ZERO RET (Zero Return).

11.2) Enter: Z

11.3) Press: SINGLE AXIS (Zero Single Axis) The Z-axis should return to it's machine zero location.

12) Remove the tool holder with edge-finder from the spindle.

12.1) Grasp the tool holder with your left hand.

12.2) Press: the TOOL RELEASE button located on the control panel. The spindle should release the tool holder. If not, make sure the machine is in the jog mode.

13) Put the tool and tool-holder back in its proper place.

Haas CNC Mill
Quick Start Up Guide
Loading Tools & Storing Tool Length Offsets

The steps outlined below are for mounting tools in a holder and loading into the Haas VF-0 or Mini Mill Machining Centers. Though intended to be used for Haas CNC machines, the concepts may be applied to other CNC machining you will find in industry. You must complete "How To Find And Store Work Zero Offsets" before proceeding further.

Note: DO NOT attempt the following procedures until they are first demonstrated by the Instructor.

1) Refer to your job tool list and collect the tools and holders required.

2) Secure the tools into the appropriate holder using the tool holder tightening fixture provided by the manufacturer.

Note: Avoid damaging the surfaces of the drill chuck arbors and tool shanks by aligning the flat recess of the tool to the set-screw of the holder before tightening.

3) Before loading tools into the machine, make sure they are placed in the correct holder and corresponding "POT" position in the tool holder (See Tool List).

4) You are now ready to load tools into the machine.

At The Control:

4.1) Confirm that the tool you are about to load is currently loaded in the spindle.

4.1.1) Press: CURNT COMDS (Current Commands).

4.1.2) View the screen: it should read "TOOL 1 IN SPINDLE." The tool number displayed should read the same as the tool your about to load. If it does, proceed to step 4.2, if it does not, continue with step 4.1.3.

4.1.3) Press: MDI (Manual Data Input) and PRGRM/CONVERS (Program).

4.1.4) Press: DELETE (optional). This is only necessary to clear the display of any old commands you do not need for this procedure. Disregard this step if there are not any commands displayed.

4.1.5) Enter: T1 M06 (input the tool number of the tool to be placed in the spindle).

4.1.6) Press: WRITE.

4.1.7) Close the machine doors.

4.1.8) Press: CYCLE START.

4.1.9) The tool should now be loaded in the spindle. Confirm this was carried out correctly by viewing the Current Commands display by pressing: CURNT COMDS. Look at the tool number displayed at the bottom of the display screen.

At The Tool Head:

4.2) Load the tool and holder in the spindle.

4.2.1) Hold the tool with your left hand.

4.2.2) Press: The TOOL RELEASE button located on the controller and hold.

4.2.3) Place the tool holder in the spindle and release the TOOL RELEASE button. Make sure it is aligned and installed correctly.

Haas CNC Mill
Quick Start Up Guide
Loading Tools & Storing Tool Length Offsets

5) You are now ready to pick-up the TLO (Tool Length Offset).

5.1) Jog the Z-axis to approximately 1" above the part or tool length offset surface. Check your set-up sheet for tool length offset surface.

5.1.1) Press: HANDLE JOG.

5.1.3) Lower the Z-axis by rotating the Jog Handle counterclockwise (CCW) in the minus (-) direction.

5.1.4) Place a 1-2-3 block on the part TLO surface and position off to the side, but not under the tool.

5.2) Jog the Z-axis to approximately 0.100" below the 1-2-3 block surface.

5.2.1) Select the .001" increment by pressing: .001 on control panel.

5.2.2) Push the 1-2-3 block lightly against the side of the tool and hold.

5.3) Raise the Z-axis by rotating the Jog Handle clockwise (CW) in the plus (+) direction until the 1-2-3 block slides under the tool and then stop.

6) You are now ready to store the Tool Length Offset (TLO).

6.1) View the tool length offsets register on control panel.

6.1.1) Press: OFFSET.

6.1.2) Press: PAGE UP or PAGE DOWN until you view the (LENGTH) register.

6.1.3) Move the cursor to the offset file number that matches the tool number in the spindle.

Note: It is extremely important that the Tool Length Offset is stored in the Correct file number. Failure to do so could result in a machine "CRASH" when the program is run.

6.2) Enter the tool length offset.

6.2.1) Press: TOOL OFSET MESUR (Tool Length Measurement). The tool length offset displayed (highlighted) should match the Z-axis position displayed at the bottom of the screen. If not, do this step again.

6.2.2) Close the machine doors.

6.2.3) Press: NEXT TOOL. The machine should raise the spindle and do a tool change. The next tool in sequence will be the next tool put in the spindle. If you are loading a tool out of sequence you will have to do this in MDI or follow the steps below:

7) Load rest of tools on the tool list and store their tool length offsets by following steps 4 through 7

8) Add –1.000 to the G54 work offset Z-axis register.

Note: Remember that the tool length offset is the distance from the bottom of the tool at machine zero to part Z zero (in this instance, the top surface of the part). Each tool offset is now set 1.000 above the top surface. To adjust for this you will make the work offset in the Z-axis –1.000.

8.1) Press: OFFSET, PAGE UP to work offset page.

8.2) Press: CURSOR up / down to G54, Z-axis..

Haas CNC Mill
Quick Start Up Guide
Loading Tools & Storing Tool Length Offsets

8.3) Enter: -1.000 and press F1.

The G54 Z register should now read -1.0000. Each tools TLO will now be added to this value.

9) Remove all tools.

9.1) Jog the current tool to machine home.

9.2) Hold the tool and press TOOL RELEASE.

9.3) Get ready to load next tool to be removed in the spindle.

9.3.1) Press: MDI (Manual Data Input)

9.3.2) CURNT COMDS
(Current Commands)

9.3.3) Press ATC REV
(Automatic Tool Change Reverse)

9.4) Remove tool.

9.5) Repeat the last two steps until all tools are removed.

ATC REV (Automatic Tool Change Reverse)

ATC FWD / REV can be used to rapidly select tools for loading or unloading. Press repeatedly until you reach the desired tool number.

Important: confirm by viewing the Current Commands display that the tool number displayed at the bottom of the screen is the same as the number corresponding with the tool being loaded.

10) Shut down the machine.

Haas CNC Simulators

Haas CNC Mill
4-Axis & 5-Axis Accessories

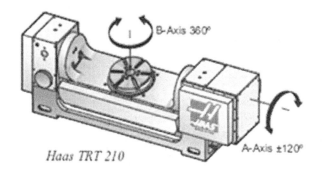

Haas TRT 210

Haas Indexer With A-Axis & B-Axis
A-Axis = 4-Axis & B-Axis = 5-Axis

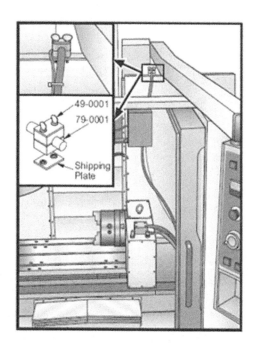

Haas rotary table model HRT 210
Installed on mill table

Haas CNC Mill
4-Axis & 5-Axis Accessories
Haas Rotary Table: Model HRT 210 (4-Axis)

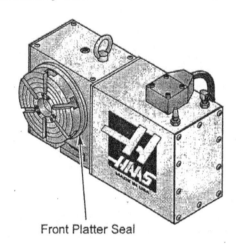

Front Platter Seal

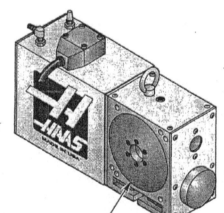

Rear Brake Seal

After use, it is important to clean the rotary table. The table should be free of any chips or grime. Clean with a chip brush and apply a coat of a rust preventative.

Do not use air-gun around front or rear seals. Chips may damage seal if blown in with an air-gun.

Haas CNC Mill
4-Axis & 5-Axis Accessories
Haas Rotary Table: Model HRT 210 (4-axis)

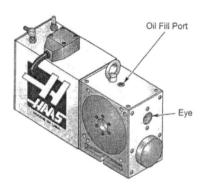

To check the lube level of the Rotary Table, view the level of lube visible in the eye with the Rotary Table stopped. The eye is located on the side panel of the Rotary Table. The lube level should reach the middle of the sight glass.

HRT210SHS – the lube level should not show more than 1/3 on the sight glass. The oil level must be to the bottom of the sight glass at a minimum.

To add lube to the Rotary Indexer, locate and remove the pipe plug from the lube fill port. This is located on the top plate above the eye (see image above).

Add Mobil SCH-630 oil until the proper level is reached. Replace the fill port bolt and tighten. Replace the oil every 2 years.

Haas A6AC Air Collet Closer

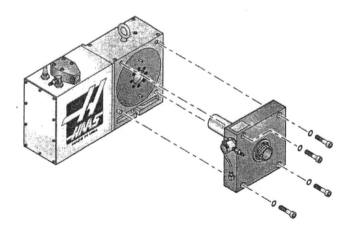

Haas CNC Mill
4-Axis & 5-Axis Accessories
Haas Rotary Table: Model HRT 210 (4-Axis)

Place the Indexer on machine. Route the cable from the table such that it avoids tool changers and table edges. Slack must be provided for the machine's movements. If the cable is cut, the motor will fail prematurely. Secure the HRT Rotary Table to machine's T-Slot table as shown below. For extra rigidity use additional Toe-Clamps.

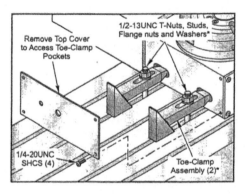

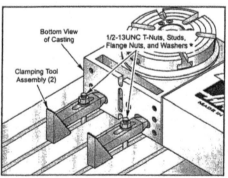

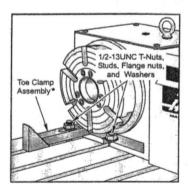

Haas CNC Mill
4-Axis Cylindrical Mapping G107

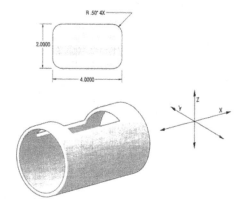

```
%
o0079 (g107 Test)
G00 G40 G49 G80 G90
G28 G91 A0
G90
G00 G54 X1.75 Y0 S5000 M03
G107 A0 Y0. R2.
(If no R or Q value, machine will use value in Setting 34)
G43 H01 Z0.25
G01 Z-0.25 F25.
G41 D01 X2. Y0.5
G03 X1.5 Y1. R0.5
G01 X-1.5
G03 X-2. Y0.5 R0.5
G01 Y-0.5
G03 X-1.5 Y-1. R0.5
G01 X1.5
G03 X2. Y-0.5 R0.5
G01 Y0.
G40 X1.75
G00 Z0.25
M09
M05
G91 G28 Z0.
G28 Y0.
G90
G107
M30
%
```

Haas CNC Mill
4-Axis Programming With Subprograms

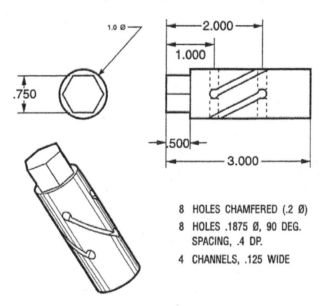

8 HOLES CHAMFERED (.2 Ø)
8 HOLES .1875 Ø, 90 DEG. SPACING, .4 DP.
4 CHANNELS, .125 WIDE

```
%
o1234
(Material is 1.0 Dia. Round stock x 3.0" long)
(Set material in collet to protrude 2.25")
(Set fixture parallel with T-slots)
(on the right side)
(X0. is the front of the material)
(Y0. is the centerline of the)
( spindle and material)
(Z0. is the top of the part)
T1 M06
G00 G90 G54 X.250 Y-.500 A0 S4500 M03
G43 H01 Z.1 M08
M98 P1235 L6
G00 G90 Z.1 M09
T2 M06
G00 G90 G54 X1.0 Y0. A0. S5000 M03
G43 H02 Z.1 M08
G82 Z-.1 F10. R.1 P300
X2.0

(continued next column)

A90.
X1.0
A180.
X2.0
A270.
X1.0
G00 G80 Z.1 M09
T3 M06
G00 G90 G54 X1.0 Y0 A0 S5000 M03
G43 H03 Z.1 M08
G83 Z-1.125 F12. R.1 Q.25
X2.0
A90.
X1.0
G00 G80 Z.1 M09
T4 M06
G00 G90 G54 X1.0 Y0. A0. S5000 M03
G43 H08 Z.1 M08
M98 P1236
G00 G90 Z.1 M09
G28 G91 Y0. Z0.
M30
%
```

Haas CNC Mill
4-Axis Programming With Subprograms

Subprogram To Mill Hex

```
%
o1235
G01 Z-.125 F50.
Y.5 F35.
G00 Z.1
G91 Y-1.0 A60.
G90
M99
%
```

The following subprograms can be written in absolute or incremental programming. Examine each program and determine which style would be faster and easier to understand and to program in the future.

Absolute:
Subprogram to mill channels

```
%
o1236
G01 Z-.25 F15.
X2.0 A90.
G00 Z.1
A180.
G01 Z-.25
X1.0 A90.
G00 Z.1
A180.
G01 Z.25
X2.0 A270.
G00 Z.1
A360.
G01 Z-.25
X1.0 A 270.
G00 Z.1
M99
%
```

Incremental:
Subprogram to mill channels

```
%
o1236
G91
G01 Z-.35 F15.
X1.0 A90.
G00 Z.35
A90.
G01 Z-.35
X-1.0 A-90.
G00 Z.35
A90.
G01 S-.35
X1.0 A90.
G00 Z.35
A90.
G01 Z.35
X-1.0 A90.
G00 G90 Z.1
M99
%
```

Haas CNC Mill
Programming 4-Axis & 5-Axis

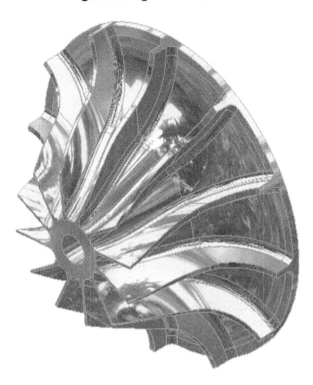

Creating Four & Five-Axis Programs:

Some 4-Axis programs are simple enough to me written manually but most 5-Axis programs are rather complex and should be written using CAD/CAM software; Mastercam or Surfcam to generate the tool paths. Complex parts can also be created with 3D solid modeling software such as Inventor, Solidworks, Pro-Engineer, or Unigraphics; exported as "step" files then imported into Mastercam or Surfcam to generate the tool paths for your CNC Mill and / or Lathe.

Simultaneous Rotation and Milling:

G94 can be used to perform simultaneous milling. The CNC relay is pulsed at the beginning of the step so that your CNC machine will proceed to the next block. The controller then executes the following L steps automatically without waiting for start commands. Normally the L count on the G94 is set to 1 and that step is followed by a step that is to be run simultaneously with a CNC mill.

Haas CNC Mill
4-Axis & 5-Axis Spiral Milling

The simultaneous rotation and milling feature of the CNC controller will permit machining of certain cam forms, spiral, and angular cuts. Spiral milling is when the spindle rotates and an axis on your mill moves at the same time. Insert a G94 into the control and the desired rotation and feed rate on the next step. The control will execute G94, this pulses the MFIN relay and allows your CNC to proceed, and the following step or steps as one step. If you wish to do more than one step, then insert the number into the L register.

By selecting a rotation feed rate and varying the mill feed rate, any spiral is possible. In order to spiral mill, you will have to calculate the feed rate for your mill so that the Haas spindle and your axis will stop at the same instant.

To calculate the feed rate for your mill you need to know.

1) The angular rotation of the spindle, this should come from the engineering drawing of the part to be made.

2) Arbitrarily select a reasonable feed rate for the spindle, a reasonable one, five-degrees (5°) per second is a good starting point.

3) The distance you wish to travel on the X-axis, this should come from the engineering drawing of the part.

For example, we wish to mill a spiral that is 72-degrees (72°) of rotation and moves 1.500 inches on the X-axis at the same time.

1) Compute the amount of time it will take the Haas index head to rotate through the angle.

$$\frac{\text{Number of Degrees}}{\text{Feedrate of Spindle}} = \text{Time to Index}$$

$$\frac{72°}{5° \text{ per sec}} = 14.40 \text{ sec for Indexing}$$

2) Now we need to compute the feed rate for the mill that will travel the X-distance in 14.40 seconds.

$$\frac{1.500 \text{ inch}}{14.4 \text{ sec}} = 0.1042$$

0.1042" per sec x 60 =
6.25 inches per minute.

6.25 inches per minute =
feed rate for mill.

Haas CNC Mill
4-Axis & 5-Axis Spiral Milling

Therefore, if you set the indexer to step 72° at a feed rate of 5° per second you will have to program your mill to travel 1.500 inches at a feed rate of 6.25 inches per minute for the spiral to be generated.

The program for the Haas control would be as follows:

Step	Step Size	Feed Rate	Loop Count	G Code
01	0	080.000	1	[94]
02	[72000]	[5.000]	1	[91]
03	0	080.000	1	[88]
04	0	080.000	1	[99]

The program for your mill would generally look like this:

N1 G00 G91 (Rapid in incremental mode)
N2 G01 F10.0 Z-1.0 (Feed down in Z-axis)
N3 M21 (To start indexing program above at step one)
N4 X-1.5 F6.25
(Index head and mill move at same time here)
N5 G00 Z1.0 (Rapid back in Z-axis)
N6 M21 (Return indexer HOME at step 3)
N7 M30

**Instructor
Mike Tatarakis
Demonstrating
To Student
How To Setup
CMM**

Haas CNC Mill
4-Axis & 5-Axis Possible Timing Problems

When the HRT executes a G94, a 250-millisecond delay is required before executing the following step. This may (it usually doesn't) cause your axis to move before the table rotates, leaving a flat spot in the cut. If this is a problem, a solution is to insert a G04 dwell (from 0 to 250-milliseconds) in your CNC after the M function to prevent axis movement. By selecting the right dwell, the HRT and your mill should start moving at the same instant. In the same manner, a problem may exist at the end of the spiral, but this can be eliminated by slightly altering the feed rate on your mill.

Don't adjust the feed rate on the Haas HRT control because your mill has a much finer feed rate adjustment than the Haas HRT control. If the undercut appears to be in the X-axis direction, then speed up slightly, your mill's feed rate by a 0.1change in feed rate. If the undercut appears in the radial direction of the spindle of the indexer, then slow down your mill's feed rate.

If the timing is off by several seconds such that your mill completes movement before the indexer completes it's movement, and you have several spiral moves one right after another, such as in retracing a spiral cut. This may cause your CNC to stop for no reason. The reason for this is your CNC will send a cycle start signal for the next cut to the Haas HRT control before it has completed its first move, thereby causing a timing hang-up.

The Haas HRT control will not accept another cycle start until it is finished with the first. If you are doing multiple moves it is very important to check your timing calculations. A way to verify if this is actually the problem is to single block your control, allowing five seconds between steps. If you can single block the control but it will not successfully run in the continuous mode, then your timing is off somewhere.

5-Axis Programming Notes

Use a tight synchronization cut across resolution of geometry in the CAD/CAM system, this will allow smooth flowing contours and a more accurate part.

Positioning the machine to an approach vector should only be done at a safe distance above or to the side of the work piece. When in the rapid mode, the axes will arrive at the programmed position at different times; the axis with shortest distance from target will arrive first, and longest distance last. A high feed-rate will force the axes to arrive at the commanded position at the same time avoiding the possibility of a crash.

G Codes:

Fifth-axis programming is not affected by the selection of inch G20 or metric G21; the A and B-axis are always programmed in degrees.

G93 inverse time must be in effect for simultaneous 5-axis motion. In G93 mode, the maximum feed-rate will include the combination of all axis motion in one block of code. The limit is set by the control and looks at encoder steps programmed for all axes in a block of code.

Haas CNC Mill
5-Axis Programming Notes

G Codes:
Limit your post processor "CAD/CAM software" if possible; the maximum speed in G93 mode is 32-degrees per minute. This will result in smoother motion, which may be necessary when fanning around tilted walls.

M Codes:
Important! It is highly recommended that the A and B-axis brakes be engaged when doing any non 5-axis cutting.

Cutting with the brakes off can cause excessive wear in the gear sets.

M Codes:
M10/M11 engages/disengages the A-axis brake

M12/M13 engages/disengages the B-axis brake

When in a 4 or 5-axis cut, the machine will pause between blocks. This pause is due to the A and/or B-axis brakes releasing. To avoid this dwell and allow for smoother program execution, program an M11 and/or M13 just before the G93. The M-codes will disengage the brakes, resulting in a smoother motion and an uninterrupted flow of motion. Remember that if the brakes are never re-engaged, they will remain off indefinitely.

Settings:
A number of settings are used to program the 4 and 5-axis.

See Settings 30, 34, and 48 for the 4-axis.

See Settings 78, 79, and 80 for the 5-axis.

Settings:
Setting 85 should be set to .0500 for 5-axis cutting. Settings lower than .0500 will move the machine closer to an exact stop and cause uneven motion. G187 can also be used in the program to slow the axes down.

Caution!
When cutting in 5-axis mode pore positioning and over-travel can occur if the tool length offset H-code is not canceled. To avoid this problem use G90, G40, H00 and G49 in your first blocks after a tool change. This problem can occur when mixing 3-axis and 5-axis programming; restarting a program or when starting a new job and the tool length offset is still in effect.

Feed-rates:
A feed-rate must be commanded for each line of 4 and/or 5-axis code. Limit the feed-rate to less than 75-IPM when drilling.

The recommended feeds for finish machining in 3-axis work should not exceed 50 to 60-IPM with at least .0500" to .0750" stock remaining for the finish operation.

Rapid moves are not allowed; rapid motions, entering and exiting holes (full retract peck-drill cycle) are not supported. When programming simultaneous 5-axis motion, less material allowance is required and higher federates may be permitted. Depending on finish allowance, length of cutter and type of profile being cut, higher feed-rates may be possible. For example, when cutting mold lines or long flowing contours, federates may exceed 100-IPM.

Haas CNC Mill
5-Axis Programming Notes

Jogging the 4 & 5-Axis:

All aspects of handle jogging for the 5-axis work as they do for the other axes. The exception is the method of selecting jog between A-axis and B-axis.

By default the "+A" and "−A" keys, when pressed, will select the A-axis for jogging. The B-axis can be selected for jogging by pressing the Shift button, and then pressing either the
"+A" or "−A" key.

EC-300: Jog mode shows A1 and A2, use "A" to jog A1 and Shift "A" to jog A2.

EC-300 Pallet and 4-axis operation:

The rotary table in the machining area will always appear, and operate, as the A-axis. The rotary axis on the pallet 1 is referred to by "A1" and the other axis, on pallet 2, by "A2".

Operation examples: To jog axis A1, enter A1 and press Hand Jog. To key jog, use the +/− A jog buttons to jog the A1-axis and +/− B buttons to jog the A2-axis. To zero return the A-axis on pallet #2, enter "A2" press Zero Single Axis.

EC-300 Mirroring Feature:

If G101 is used to mirror the A-axis, then mirroring is turned on for both A-axes. When pallet #1 is in the machining area, A1-MIR will be displayed at the bottom of the screen. When pallet #2 is in the machine, A2-MIR will be displayed. The behavior of the Mirror Settings is different. If Setting 48 Mirror Image A-axis is ON only the A-axis on pallet #1 is mirrored and the message A1-MIR is displayed.

If Setting 80 (parameter 315, bit 20 MAP 4TH AXIS is a 1, the name for setting 80 is the same as that for Setting 48, i.e. Mirror Image A-Axis) is ON mirroring will be turned on for the A-axis on pallet #2. When pallet #2 is inside of the mill, A2-MIR will be displayed.

Crash Recovery Procedure:

If the machine crashes while cutting a 5-axis part, it can often be difficult to clear the tool away from the part due to the angles involved. Do Not immediately press the Recover button or turn the power off. To recover from a crash in which the spindle is stopped while the tool is still in a cut, retract the spindle using the Vector Jog feature. To do this press the letter "V" on the keypad press "Handle Jog", and use the jog handle to move along that axis. This feature will allow motion along any axis determined by A and/or B-axis.

The Vector jog feature is intended to allow the operator to clear the cutting tool from the part in an extreme situation as a result of a crash or an alarm condition.

G28 is not available in Vector Jog mode; it is only available for the X, Y, Z, A, and B when selecting single axis.

If there was a loss of power during a cut Vector Jog will not work as the control requires a reference position. Other means of clearing the tool from the part will be necessary.

If the tool is not in a cut when it is crashed, press the Recover button and answer the questions that appear on the screen. When the Recover button is pressed, the spindle head will move the A, B, and Z-axis simultaneously, in order to retract the tool. If the tool is in a cut at an angle, it will crash when this key is pressed.

Haas CNC Mill
5-Axis G-Codes

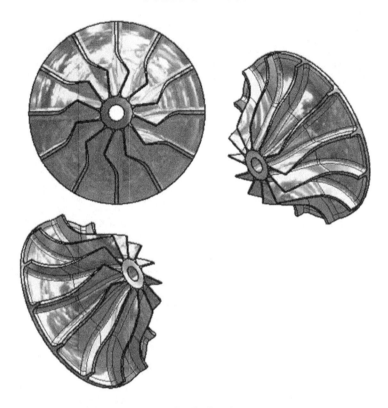

5-Axis G-Codes

Code, Group, Description

G143, 08, ** 5-Axis Tool Length Comp + (X,Y,Z,A,B,H) (Setting 15,117)
G153, 09, ** 5-Axis HSP Drill (X,Y,A,B,Z,I,J,K,Q,P,E,L,F,) (Setting 22)
G155, 09, ** 5-Axis Reverse Tapping Canned Cycle (X,Y,A,B,Z,J,E,L,F)
G161, 09, ** 5-Axis Drill Canned Cycle (X,Y,A,B,Z,E,L,F)
G162, 09, ** 5-AxisSpot Drill Canned Cycle (X,Y,A,B,Z,J,E,L,F)
G163, 09, ** 5-Axis Peck Drill Canned Cycle (X,Y,A,B,Z,I,J,K,Q,P,E,L,F) (Setting 22)
G164, 09, ** 5-Axis Tapping Canned Cycle (X,Y,A,B,Z,J,E,L,F)
G165, 09, ** 5-Axis Bore In, Bore Out Canned Cycle (X,Y,A,B,Z,E,L,F)
G166, 09, ** 5-Axis Bore In, Stop, Rapid Out Canned Cycle, (,Y,A,B,Z,E,L,F)
G169, 09, ** 5-Axis Bore In, Dwell, Bore Out Canned Cycle, (X,Y,A,B,Z,P,E,L,F)

Haas CNC Mill
G153, 5-Axis High Speed Peck Drilling With I, J & K

- E Specifies the distance from the start position to the bottom of the hole.
- F Feed-rate in inches or (mm) per minute
- I Size of first cutting depth (must be a positive value).
- J Amount to reduce cutting depth each pass (must be a positive value).
- K Minimum depth of cut (must be a positive value).
- L Number of repeats
- P Pause at end of last peck, in seconds.
- A A-axis tool starting position
- B B-axis tool starting position
- X X-axis tool starting position
- Y Y-axis tool starting position
- Z Z-axis tool starting position

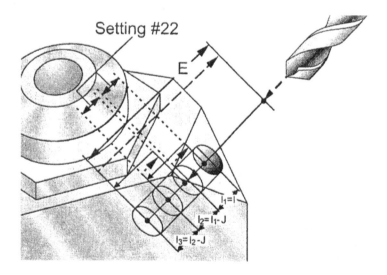

This is a high-speed peck cycle where the retract distance is set by setting 22. If I, J, and K are specified, a different operating mode is selected. The first pass will cut in by amount I, each succeeding cut will be reduced by amount J, and the minimum cutting depth is K. If P is used, the tool will pause at the bottom of the hole for that amount of time. Note that the same dwell time applies to all subsequent blocks that do not specify a dwell time.

A specific X, Y, Z, A, B, position must be programmed before the canned cycle is commanded. This position is used as the initial start position.

Haas CNC Mill
G153, 5-Axis High Speed Peck Drilling with K & Q

- E Specifies the distance from the start position to the bottom of the hole.
- F Feed-rate in inches or (mm) per minute
- K Minimum depth of cut (must be a positive value).
- L Number of repeats
- P Pause at end of last peck, in seconds.
- Q The cut-in value (must be a positive value).
- A A-axis tool starting position
- B B-axis tool starting position
- X X-axis tool starting position
- Y Y-axis tool starting position
- Z Z-axis tool starting position

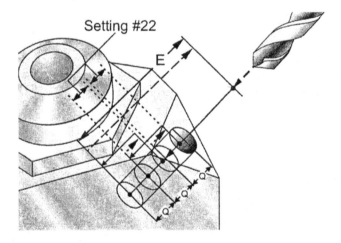

A specific X, Y, Z, A, B, position must be programmed before the canned cycle is commanded. This position is used as the initial start position.

Haas CNC Mill
G155, 5-Axis Reverse Tapping Canned Cycle

G155 only performs floating taps. G174 (CCW) is available for 5-axis reverse rigid tapping.

E   Specifies the distance from the start position to the bottom of the hole.
F   Feed-rate in inches (mm) per minute
L   Number of repeats
A   A-axis tool starting position
B   B-axis tool starting position
X   X-axis tool starting position
Y   Y-axis tool starting position
Z   Z-axis tool starting position
S   Spindle speed

You do not need to start the spindle CCW before this canned cycle. The control does this automatically.

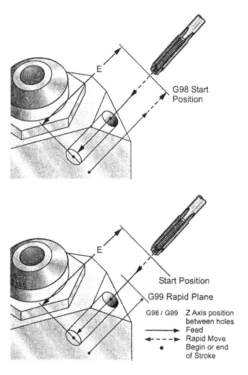

A specific X, Y, Z, A, B, position must be programmed before the canned cycle is commanded. This position is used as the initial start position.

Haas CNC Mill
G161, 5-Axis Drill Canned Cycle

E Specifies the distance from the start position to the bottom of the hole.
F Feed-rate in inches (mm) per minute
L Number of repeats
A A-axis tool starting position
B B-axis tool starting position
X X-axis tool starting position
Y Y-axis tool starting position
Z Z-axis tool starting position

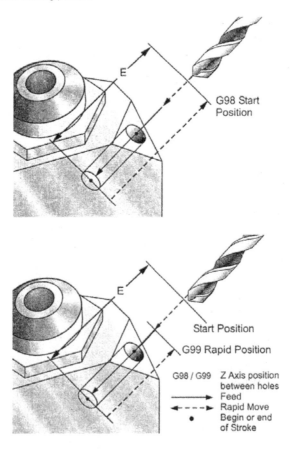

A specific X, Y, Z, A, B, position must be programmed before the canned cycle is commanded. This position is used as the initial start position.

Haas CNC Mill
G162, 5-Axis Spot Drill Canned Cycle

E Specifies the distance from the start position to the bottom of the hole.
F Feed-rate in inches (mm) per minute
L Number of repeats
P The dwell time at the bottom of the hole
A A-axis tool starting position
B B-axis tool starting position
X X-axis tool starting position
Y Y-axis tool starting position
Z Z-axis tool starting position

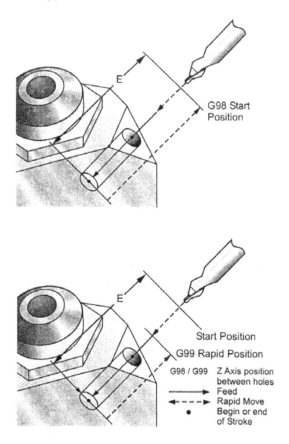

A specific X, Y, Z, A, B, position must be programmed before the canned cycle is commanded. This position is used as the initial start position.

Haas CNC Mill
C163, 5-Axis Normal Peck Drilling Canned Cycle

- E Specifies the distance from the start position to the bottom of the hole.
- F Feed-rate in inches or (mm) per minute
- I Optional size of first cutting depth
- J Optional amount to reduce cutting depth each pass
- K Optional minimum depth of cut
- L Number of repeats
- P Pause at end of last peck, in seconds
- Q The cut-in value, always incremental
- A A-axis tool starting position
- B B-axis tool starting position
- X X-axis tool starting position
- Y Y-axis tool starting position
- Z Z-axis tool starting position

A specific X, Y, Z, A, B, position must be programmed before the canned cycle is commanded. This position is used as the initial start position.

If I, J, and K, are specified the first will cut in by amount I, each succeeding cut will be reduced by amount J, and the minimum cutting depth is K.

If a P value is used the tool will pause at the bottom of the hole after the last peck for the amount of time. The following example will peck several times and dwell for one and half seconds at the end.

G163 Z-0.62 F15. R0.1 Q0.175 P1.5

Note that the same dwell time applies to all subsequent blocks that do not specify a dwell time.

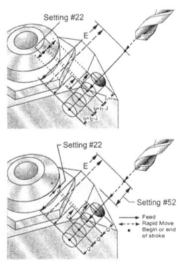

Setting 52 also changes the way G163 works when it returns to the start position. Usually the R plane is set well above the cut to ensure that the peck motion allows the chips to get out of the hole. This wastes time as the drill starts by drilling "empty" space. If Setting 52 is set to the distance required to clear the chips, the start position can be much closer to the part being drilled. When the chip-clearing move to the start position occurs, the Z-axis will be moved above the start position by the amount given in this setting.

Haas CNC Mill
G164, 5-Axis Tapping Canned Cycle

G164 only performs floating taps. G184 (CW) is available for 5-axis rigid tapping.

- E   Specifies the distance from the start position to the bottom of the hole.
- F   Feed-rate in inches (mm) per minute
- L   Number of repeats
- A   A-axis tool starting position
- B   B-axis tool starting position
- X   X-axis tool starting position
- Y   Y-axis tool starting position
- Z   Z-axis tool starting position
- S   Spindle speed

You do not need to start the spindle CW before this canned cycle. The control does this automatically.

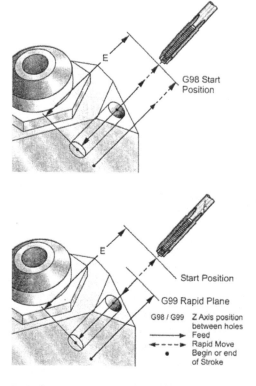

A specific X, Y, Z, A, B, position must be programmed before the canned cycle is commanded. This position is used as the initial start position.

Haas CNC Mill
G165, 5-Axis boring Canned Cycle

E   Specifies the distance from the start position to the bottom of the hole.
F   Feed-rate in inches (mm) per minute
L   Number of repeats
A   A-axis tool starting position
B   B-axis tool starting position
X   X-axis tool starting position
Y   Y-axis tool starting position
Z   Z-axis tool starting position

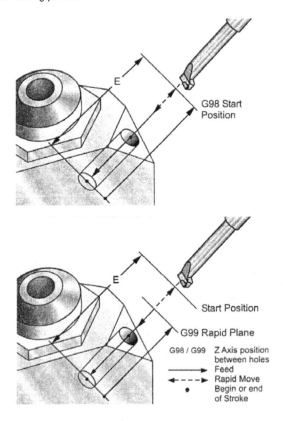

A specific X, Y, Z, A, B, position must be programmed before the canned cycle is commanded. This position is used as the initial start position.

Haas CNC Mill
G166, 5-Axis Bore & Stop Canned Cycle

E Specifies the distance from the start position to the bottom of the hole.
F Feed-rate in inches (mm) per minute
L Number of repeats
A A-axis tool starting position
B B-axis tool starting position
X X-axis tool starting position
Y Y-axis tool starting position
Z Z-axis tool starting position

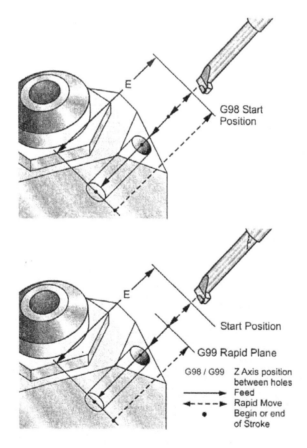

A specific X, Y, Z, A, B, position must be programmed before the canned cycle is commanded. This position is used as the initial start position.

Haas CNC Mill
G169, 5-Axis Bore & Dwell Canned Cycle

E Specifies the distance from the start position to the bottom of the hole.
F Feed-rate in inches (mm) per minute
L Number of repeats
P The dwell time at the bottom of the hole
A A-axis tool starting position
B B-axis tool starting position
X X-axis tool starting position
Y Y-axis tool starting position
Z Z-axis tool starting position

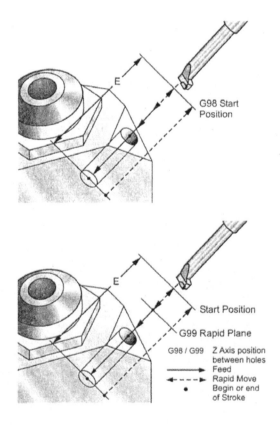

A specific X, Y, Z, A, B, position must be programmed before the canned cycle is commanded. This position is used as the initial start position.

Haas CNC Mill
G174 CCW Non-Vertical Rigid Tapping
G184 CW Non-Vertical Rigid Tapping

F   Feed-rate in inches (mm) per minute
A   A-axis tool starting position
B   B-axis tool starting position
X   X-axis tool starting position
Y   Y-axis tool starting position
Z   Z-axis tool starting position
S   Spindle Speed

You do not need to start the spindle before this canned cycle. The control does this automatically.

A specific X, Y, Z, A, B, position must be programmed before the canned cycle is commanded. This position is used as the initial start position.

This G code is used to perform rigid tapping for non-vertical holes. It may be used with a right-angle head to perform rigid tapping in the X or Y-axis on a three-axis mill, or to perform rigid tapping along an arbitrary angle with a five-axis mill. The ratio between the federate and spindle speed must be precisely the thread pitch being cut.

Instructor, Jim Mori in the tool crib

De Anza, Manual Machine Lab

Manual Lathe

Manual Lathe

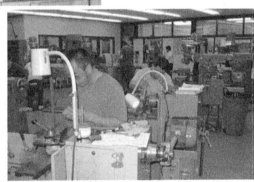

Inspection Surface Plate

Kalamazoo Cut Off Saw

De Anza, Manual Machine Lab

Instructor Jim Mori Demonstrates To Student How To Setup Bridgeport Mill

Manual Lathe

Bridgeport Mill

De Anza, CNC Lab Projects
Rick Jow's MCNC 56 Special Project

Manual to CNC Mill Conversion

All motor mounts, lead screw drives, belt guards, etc., designed and machined by Rick Jow.

Instructor Chris Newell's MCNC 75B Class Project
Low Temperature Differential Stirling Heat Engines

EAPPRENTICE.NET

Mastercam Classes On Line

**Mastercam Video
Author
Mastercam Instructor**

**Haas CNC Mill
Fourth & Fifth Axis
Programming
&
Haas CNC Lathe
Live Tooling
Programming**

Derek Goodwin
Has worked in the CNC machining industry for over 30 years. He served a four-year apprenticeship in the U.K. while attending Carlett Part college of Technology, attained the Engineering Industry Training Board certificate of engineering craftsmanship and City and Guilds Advanced certificate of Mechanical Engineering craft studies. Parts 1, 2 and 3 After returning to the U.S. he founded Goodwin Design and Mfg. in the heart of Silicon Valley, providing precision machined components to the High Tech industry for over 15 years. After selling his company in 2004, Derek joined the International Design firm IDEO, where he leads the Prototyping team. He also teaches CAD/CAM Programming at De Anza College in Cupertino, CA. and is the founder of eapprentice.net where he teaches Mastercam programming on the internet.

**Mastercam
5 – Axis
CNC Project**

Haas CNC Lathes

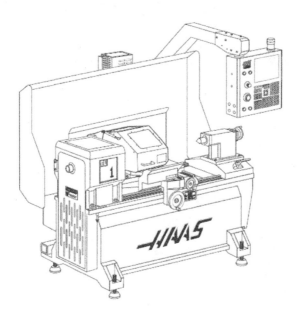

Haas Toolroom Lathe

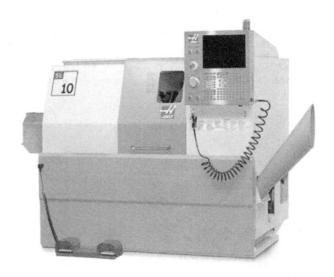

Haas SL-10 Lathe

Haas CNC Lathe Coordinates

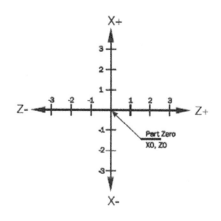

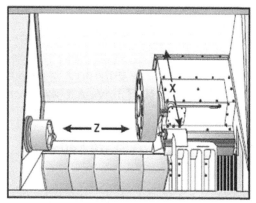

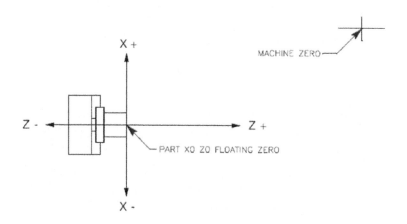

Haas CNC Lathe Coordinate System

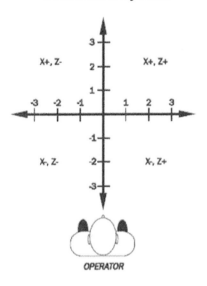

**CNC Lathes have only an X & Z-Axis
They DO NOT have a Y-axis**

Haas CNC SL-10 Lathe

Haas CNC Lathe
Lathe Introduction

This manual provides basic principles of programming that's necessary to begin programming the Haas CNC Lathe.

The CNC (Computerized Numerical Control) machine, the tool is controlled by a computer and is programmed with machine code system that enables it to be operated with minimal supervision and with a great deal of repeatability.

The same principles used in operating a manual machine are used in programming a CNC machine. The main difference is that instead of cranking handles to position a slide to a certain point, the dimension is stored in the memory of the machine control "once." The control will then move the machine to these positions each time the program is run.

In order to operate and program a CNC controlled machine, a basic understanding of machining practices an a working knowledge of mathematics is necessary.

It is also important to become familiar with the control console and the placement of the keys, switches, displays, etc., that are pertinent to the operation of the machine.

This Haas lathe manual can be used for both operators and programmers. It is meant as a supplementary teaching aid for the users of Haas Lathes. The information in this manual may apply in whole or in part to the programming of other CNC machines.

This manual is intended to give a basic understanding of CNC programming and it's applications. It is not intended as an in-depth study of all ranges of machine use, but as an overview of common and potential situations facing CNC programmers. Much more training and information is necessary before attempting to program on the machine.

For a complete explanation and in-depth description, refer to the Programming and Operation Manual that is supplied with your Hass Lathe.

Haas CNC SL-20 Lathe

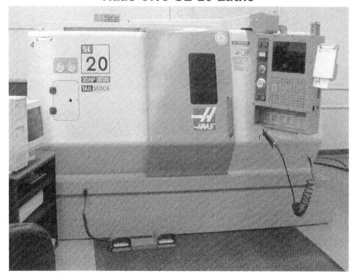

Haas CNC Lathe
Lathe Programming With Codes

The definition of a part program for any CNC consists of movements of the tool, and speed changes to the spindle RPM. It also contains auxiliary command functions such as tool changes, coolant on or off commands, or external M code commands.

Tool movements consist of rapid positioning commands, straight line moves or movement along an arc of the tool at a controlled speed.

The Haas lathe has two (2) linear axes defined as X-axis, and Z-axis. The X axis moves the tool turret toward and away from the spindle center line, while the Z axis moves the tool turret along the spindle axis. The machine-zero position is where the tool is at the upper right corner of the work cell farthest away from the spindle axis.

Motion in the X-axis will move the turret toward the spindle centerline with negative values and away from spindle center with positive values. Motion in the Z axis will move the tool toward the spindle chuck with negative values and away from the machine chuck with positive values.

A program is written as a set of instructions given in the order they are to be performed. The instructions, if given in English, might look like this:

Line #1 = Select Cutting Tool

Line #2 = Turn Spindle On And Select The RPM

Line #3 = Rapid To The Starting Position Of The Part

Line #4 = Turn Coolant On

Line #5 = Choose Proper Feed-rate And Make The Cut (s)

Line #6 = Turn The Spindle And Coolant Off.

Line #7 = Return To Clearance Position To Select Another Tool and so on.

But our machine control understands only these messages when given in machine code, also referred to as G and M codes programming. Before considering the meaning and the use of codes, it is helpful to lay down a few guidelines.

Haas CNC Lathe
Lathe Program Format

There is no positional requirement for the address codes. They may be placed in any order within the block. Each individual can format their programs in many different ways. But, program format or program style is an important part of CNC machining. There are some program command formats that can be moved around, and some commands need to be a certain way, and there are some standard program rules that are just good to follow. The point is that a program needs to have an organized program format that's consistent and efficient so that any CNC machinist in your shop can understand it.

Some standard program rules to consider are:

Program X, and Z in alphabetical order on any block. The machine will read Z or X in any order, but we want to be consistent. If both X and Z are on a command line in a program, they should be listed together and in order. Write X first, and Z second.

You can put G and M codes anywhere on a line of code. But, in the beginning when N/C programming was being developed G codes had to be in the beginning of a program line and M codes had to be at the end. And this rule, a lot of people still follow and is a good standard to continue.

Some CNC machines allow you to write more than one M code per line of code and some won't. On the Haas, only one M code may be programmed per block and all M codes are activated or cause an action to occur after everything else on the line has been executed.

Program format is a series and sequence of commands that a machine may accept and execute. Program format is the order in which the machine code is listed in the program that consist of command words. Command words begin with a single letter and then numbers for each word. If it has a plus (+) value, no sign is needed. If it has a minus value, it must be entered with a minus (-) sign. If a command word is only a number and not a value, then no sign or decimal point is entered with that command. Program format defines the language of the machine tool.

. . .

G82 Z-0.2 P0.3 R0.1 F0.003
G80 G00 Z1. M09
G28
M01
N4 (Drill .312 Dia. X 1.5 Depth)
G28
T404 (5/16 Dia. Drill)
G97 S2400 M03
G54 G00 X0. Z1. M08
G83 Z-1.5 Q0.3 R0.1 F0.006
G80 G00 Z1. M09
G28 M01

. . .

Haas CNC Lathe
Lathe Definitions Within The Format

1) Character: A single alphanumeric character value or the "+" and "-" sign.

2) Word: A series of characters defining a single function such as, G codes, M codes, an "X" axis moves, or "F" federate. A letter is the first character of a word for each of the different commands. There may be a distance and direction defined for a word in a program. The distance and direction in a word is made up of a value, with a plus (+) or minus (-) sign. A plus (+) value is recognized if no sign is given in a word.

3) Block: series of words defining a single instruction. An instruction may consist of a single linear motion, a circular motion or canned cycle, plus additional information such as a federate or miscellaneous command (M-Codes).

4) Positive Signs: If the value following an address letter command such as A, B, C, I, J, K, R, U, V, W, X, Y, Z, is positive, the plus sign need not be programmed in. If it has a minus value it must be programmed in with a minus (-) sign.

5) Leading Zeros: if the digits preceding a number are zero, they need not be programmed in.
The Haas control will automatically enter in the leading zeros.
Example: G0 for G00 and M1 for M01.
Trailing zeros must be programmed.
Example: M30 not M3, G70 not G7.

6) Model Commands: Codes that are active for more than the line in which they are issued are called "Model" commands. Rapid traverse, federate moves, and canned cycles are all examples of modal commands.
A "Non-Modal command which once called, are effective only in the calling block, and are then immediately forgotten by the control.

7) Preparatory Functions: "G" codes use the information contained on the line to make the machine tool do specific operations, such as:
a) Move the tool at rapid traverse.

b) Move the tool at a feed-rate along a straight line.

d) Move the tool along an arc at a feed-rate in a clockwise direction.

e) Move the tool along an arc at a feed-rate in a counterclockwise direction.

f) Move the tool through a series of repetitive operations controlled by "fixed cycles" such as, spot drilling, drilling, boring, and tapping.

8) Miscellaneous Functions: "M" codes are effective or cause an action to occur at the end of the block, and only one "M" code is allowed in each block of a program.

9) Sequence Numbers: N1 through N99999 in a program are only used to locate and identify a line or block and its relative position within a CNC program.

A program can be with or without Sequence Numbers. The only function of Sequence Numbers is to locate a certain block or line within a CNC program.

The machining cycles G70, G71, G72 and G73 require the use of sequence numbers to call up specified blocks in a program that in it define the part geometry to rough and finish.

Haas CNC Lathe
Lathe Program First Couple Of Lines

The FIRST LINE or block in a program should be a return to machine zero (using G28 or G51 codes). Any tool change should be after a return to machine zero or a tool change location. Although this not necessary it is a good safety measure.

The SECOND LINE of code should apply to any appropriate tool selections and tool geometry offsets or tool shifts.

The THIRD LINE may optionally contain a spindle speed maximum for the tool being used.

The FOURTH LINE or block should cancel any constant surface speed mode (G97). And it should specify a constant spindle speed command (Snnnn) along with a spindle ON clockwise command (M03).

The FIFTH LINE should contain a work offset (G54), a preparatory code (G00) for rapid command with an X and Z location for positioning the turret, and turn on the coolant (M08).

The SIXTH LINE may choose to, optionally specify a constant surface speed with (G96) and a surface feet per minute (SFM) defined with a (Snnnn) command.

An example of the programs startup lines might look like this:

With constant surface speed.

N11 G28
N12 T101 (O.D. Tool x .031 TNR)
N13 G50 S2800
N14 G97 S650 M03
N15 G54 G00 X1.85 Z1. M08
N16 G96 S315

Without constant surface speed.

N21 G28
N22 T101 (.750 Dia. Drill)
N23 G97 S1600 M03
N24 G54 G00 X0. Z1. M08

Note:
All the necessary codes for each operation are listed in the following pages. This tool startup format is a good example and defines a commonly used program style.

Haas CNC Lathe
Lathe Safe Startup Line

G18 G20 G40 G54 G80 G97 G99

Do you need a safe startup line to be sure all the commands are canceled before starting a program? Many programs have a G code default line (Or the CAD / CAM programming system may output a default line) at the beginning. To be sure machine control is in a safe start condition. Here are the conditions on a Haas control, to help you decide if you need a safe start line at the beginning of a program.

G18 Circular Motion ZX Plane Selection:
The G18 is the default condition on a Haas lathe, and is the only one available in the X and Z-axis to do an arc. If you try an arc in the G17 XY plain, or the G19 YZ plain, the machine would stop and give you an alarm. There is no need to program a G18, it is already active when you power up.

G20 Inch / G21 Metric Dimensioning:
The G codes G20 and G21 are used to select between inch and metric. On the Haas control, the G20 (inch) and G21 (mm) codes are used to make sure the inch / metric setting (Setting 9) is set correctly for that program.

If you're not switching between inch and metric very much, then why keep putting in these G codes at the start of every program? If you did have the wrong condition in Setting 9, graphics display will display whatever Setting 9 is set at. You'll easily see the error when running a program in graphics.

G54 Work Offset Command:
Work offsets on a CNC lathe, are not used like they are on a CNC mill. Many lathe users don't even have values in their work offsets, because all the offsets that are needed on a lathe, for most users, are entered in the Tool Geometry display.

The only time someone might use a work offset on a lathe, is to shift all the tools in Z-axis the same distance.

Example: lets say you want to move all your tools -0.015 in the Z axis, to make the parts that much shorter. Instead of changing all tool offsets. You could shift all the tools in -0.015 by changing the work offset. Just be sure to take this offset shift out when you're done using it.

Work Zero Offset:

G code	(X)	(Z)	(B)
G52	0.	0.	0.
G54	0.	-0.015	0.
G55	0.	0.	0.

Another main reason to use a work offset on a lathe, is when you're touching off your tools on a tool probe. The X axis value is usually always good to the center of the spindle using the value touched off in the X axis on a tool probe along with the value in Setting 59 or 60. The problem is the distance from the edge of the tool probe in the Z axis to the face of the part. So compensate for that difference in the Work Zero Offset, that has in it, the extra distance from the edge of the tool probe to the face of the part, in the Z-axis.

Be aware that even if you're not using a work-offset command in a program, G54 is still active as a default G code. This is why you see G54 in the program examples of this book, as a reminder that this work offset is active whether you program it in or not. If you're never using work offsets, you may choose to leave this G54 command out.

Haas CNC Lathe
Lathe Safe Startup Line

G40 Cancels Tool Nose Compensation:
You may see a G40 in the beginning of every program. And some even put it in at the beginning of every tool; to be sure cutter compensation is off, before they start a program.

You should always cancel cutter compensation (G40) when you're done using it. If you forget to cancel cutter compensation on a lathe, and you run a program in graphics, you'll get a 349 PROG STOP W/O CANCEL CUTTER COMP alarm. To let you know, that you ended a program without canceling cutter compensation. Pressing RESET or POWER OFF will also cancel cutter compensation. Because of these conditions that cancel cutter compensation, you don't need to put a G40 at the beginning of every program.

G80 Cancels Canned Cycles:
If you forget to cancel a canned cycle with a G80, RESET, G00, G01, M30 and Power Down will also cancel any active canned cycle.

G97 Constant Non-Varying Spindle Speed: For safety reasons you should program in a G97 at the beginning and end of every tool cycle, with a spindle speed.

G99 Feed Per Revolution:
This command changes how the "F" address is interpreted. The F command indicates inches per spindle revolution when Setting 9 is set to INCH.

If Setting 9 is set to METRIC, F indicates millimeters per revolution. G99 and G98 are model commands. G99 is the default command, and the one you'll usually want. In certain situations you may choose G98 for Feed Per Minute, and then switch back to G99. but most customers, will already be in G99 and never switch to G98, so there's usually no need to have it in a safe start line. If a mistake is made, using the wrong feed command, you'll easily see it when it happens, when running the program on the machine.

Example: If you program a feed of F0.005 to feed per revolution, and you accidentally program in a G98 (Feed Per Minute), then it would take forever to feed .005 a minute. If you programmed F10. for feed per minute (G98), and you were accidentally in Feed Per Revolution (G99), and you had the spindle on, lets say at a low speed of S200. the calculated federate for this would be 2000 Inches Per Minute, which is above the machine maximum feed rate. The machine would stop and give you an alarm.

There might be good reasons why you want a safe start line at the beginning of every program, and some programmers like having it at the beginning of every tool, as a safety precaution. You may have other machines running these same programs that may need safe startup lines. But for many, it may not be necessary. Because of these reasons listed above. It's up to the programmer to decide if they want safe startup lines in their programs.

Haas CNC Lathe
Lathe Program Structure

A CNC part program consists of one or more blocks of commands. When viewing the program, a block is the same as a line of text. Blocks shown on the control display are always terminated by the ";" symbol which is called an EOB (End Of Block). Blocks are made up of alphabetical address codes; which are always an alphabetical character followed by a numeric value. For instance, the specification to move the X-axis would be a number proceeded by the X symbol.

Programs must begin and end with a percent (%) sign. After the first percent (%) sign, the program must have a program number beginning with the letter O (not zero) and then the number that defines that program (four digit number for older machines and five digit number for newer machines). Those program numbers are used to identify and select a main program to be run, or as a sub-program called up by the main program. The % signs will "not" be seen on the control, but they must be in the program when you load it into the control. And they will be seen when you download a program from the machine. The % signs are automatically entered in for you, if you enter the program in on the Haas control.

A program may also contain a "/" symbol. The "/" symbol, sometimes called a forward slash, is used to define an optional block. If a block contains this symbol, any information that follows the forward slash in a program block, will be ignored when the "Block Delete" button is selected when running a program.

It is common to begin each tool in a part program with preparatory codes, turning on commands associated for the tool, and then ending by returning to machine home, or a safe location, to position for a tool change. There might be a number of commands that are repeated throughout the program. This is done for safety to insure that the proper commands are attained if the operator has to begin, at the start of a tool in the middle of a program, in the event of tool breakage, to rerun a tool, or finish a part after powering up the machine. This is a common programming practice.

On the following page is a sample program as it would appear on the machine control display.

The program will rough and finish turn and face for a part with two diameters, along with drilling a 3/8" Dia. X 1.000 deep hole.

%	Programs begin & end with a %
o00018 (lathe program name)	Letter "O" and five digit program number.
N1 (rough O.D.	First operation
G28	Return to machine zero for tool change.
T101 (O.D. tool x .031 TNR)	Select tool #1 with offset #1.
G50 S2600	Set spindle speed max. 2600 rpm.
G97 S415 M03	Cancel CSS, 415 rpm, ON forward.
G54 G00 X3.6 Z0.1 M08	Work offset, rapid X, Z axis, coolant ON.
G96 S390	CSS on at 390 SFM, coolant ON.
G00 Z0.005	Rapid to .005 from end of part.
G01 X-0.063 F0.005	Rough face end of part.
G00 X3.6 Z0.1	Rapid to start point above part
G71 P10 Q20 U0.01 W0.005 D0.1 F0.01	Rough turning G71 canned cycle.

Haas CNC Lathe
Lathe Program Structure

N10 G42 G00 X0.82	N10 starting block called by G71 "P10"
G01 Z0.0 F0.004	Line defining part geometry to rough out.
X0.9	Line defining part geometry to rough out.
G03 X1.0 Z-0.05 R0.05	Line defining part geometry to rough out.
/G01 Z-1.75	Line defining part geometry to rough out.
/X1.75	Line defining part geometry to rough out.
/G03 X2.25 Z-2. R.25	Line defining part geometry to rough out.
G01 Z-3.25 F0.004	Line defining part geometry to rough out.
X2.94	Line defining part geometry to rough out.
X3. Z-3.28	Line defining part geometry to rough out.
Z-4.1	Line defining part geometry to rough out.
N20 G40 X3.6	N10 Ending block called by G71 "Q20"
G97 S414 M09	Cancel CSS, coolant off.
G28	Return to machine zero for tool change.
M01	Optional program stop.
N2 (finish O.D.)	Second operation
G28	Return to machine zero for tool change.
T202 (O.D. tool x .031 TNR)	Select tool #2 with offset #2
G50 S2600	Spindle speed max. 2600 rpm.
G97 S1354 M03	Cancel CSS, 1350 rpm, ON forward
G54 G00 X1.1 Z0.1 M08	Work offset, rapid X, Z axis, coolant ON.
G96 S390	CSS on at 390 SFM.
G00 Z0.	Rapid to end of part.
G01 X-032 F.003	Finish face end of part.
G00 X3.6 Z0.1	Rapid to start point above part.
G70 P10 Q20	Finishing cycle calling N10 & N20.
G97 S414	Cancel CSS, 475 rpm spindle speed.
G00 Z1.0 M09	Rapid z axis, coolant off.
G28	Return to machine zero for tool change.
M01	Optional program stop.
N3 (drill .375 Dia. X 1.0 depth)	Third operation.
G28	Return to machine zero for tool change.
T303 (3/8" Dia. drill)	Select tool #3 with offset #3.
G97 S1986 M03	Cancel CSS, 1950 rpm, ON forward.
G54 G00 X0. Z1. M08	Work offset, rapid X, Z axis, coolant ON.
G83 Z-2.5 Q0.3 R0.1 F0.005	Deep hole peck drill 2.5 deep, 0.3 peck.
G80 G00 Z0.1 M09	Cancel canned cycle, rapid Zaxis,
G28	Return to machine zero for tool change.
T100	Select tool #1 for the next part.
M30	Stop program.
%	Programs begin & end with a %

Haas CNC Lathe
Often Used Preparatory "G" & "M" Codes

Often Used Preparatory "G" Codes

G00 – Rapid traverse motion; Used for non-cutting moves of the machine in positioning quick to a location to be machined, or rapid away after program cuts have been performed. Maximum rapid motion (IPM) of a Hass machine will vary on machine model.

G01 – Linear interpolation motion; Used fro actual machining and metal removal. Governed by a programmed feed-rate in inches (or mm) per revolution (G99).
Maximum feed-rate (IPM) of a Hass machine will vary on machine model.
(Inch Per Minute = RPM x Inch Per Rev.)

G02 – Circular interpolation, CW.

G03 – Circular interpolation, CCW.

G28 – Machine Home (Rapid Traverse)

G40 – Tool Nose Compensation Cancel.

G41 – Tool Nose Compensation LEFT of the programmed path.

G42 – Tool Nose Compensation RIGHT of the programmed path.

G50 – Spindle Speed Max RPM Limit.

G70 – Finishing Cycle

G71 – O.D. / I.D. Stock Removal Cycle.

G72 – End Face Stock Removal Cycle.

G71 – O.D. / I.D. Thread Cutting Cycle.

G80 – Cancel Canned Cycle.

G81 – Drill Canned Cycle.

G83 – Peck Drill Canned Cycle.

G84 – Tapping Canned Cycle.

G96 – Constant Surface Speed ON.

G96 – Constant Surface Speed Cancel.

Often Used Preparatory "G" Codes

G98 – Feed Per Minute.

G99 – Feed Per Revolution.

Often Used Preparatory "M" Codes

M00 – The M00 code is used for a program stop command on the machine. It stops the spindle, turns off coolant and stops look-a-head processing. Pressing CYCLE START again will continue the program on the next block of the program.

M01 – The code is used for an optional program stop command. Pressing the OPT STOP key on the control panel signals the machine to perform a stop command when the control reads an M01 command. It will then perform like an M00.

M03 – Starts the Spindle FORWARD. Must have a spindle speed defined.

M04 – Starts the Spindle REVERSE. Must have a spindle speed defined.

M05 – STOPS the Spindle.

M08 – Coolant ON command.

M09 – Coolant OFF command.

M30 – Program End and Reset to the beginning of program.

M97 – Local Subroutine Call.

M98 – Subprogram Call.

M99 – Subprogram Return (M98) or Subroutine Return (M97), or a Program Loop.

Note:
Only one "M" code can be used per line. And the "M" codes will be the last command executed in a line, regardless of where it's located in the line.

Haas CNC Lathe: G-Codes

1) G codes come in groups. Each group of G codes will have a specific group number.

2) A G code from the same group can be replaced by another G code in the same group. By doing this the programmer establishes modes of operation. The universal rule here is that codes from the same group cannot be used more than once on the same line.

3) There are Modal G codes which once established, remain effective until replaced with another G code from the same group.

4) There are Non-Modal G codes (Group 00) which once called, are effective only in the calling block, and are immediately forgotten by the control.

Each G code is a part of a group of G codes. The Group 0 codes are non-model; that is, they specify a function applicable to this block only and do not affect other blocks.

The other groups are model and the specification of one code in the group cancels the previous code applicable from that group. A model G code applies to all subsequent blocks so those blocks do not need to re-specify the same G code.

There is also one case where the Group 01 G co0des will cancel the group 9 (canned cycles) codes. If a canned cycle is active (G81 through G89), the use of G00 or G01 will cancel the canned cycle

The rules above govern the use of the G codes used for programming the Haas Lathe. The concept of grouping codes and the rules that apply will have to be remembered to effectively program the Haas Lathe. The following is a list of Haas G codes. If there's a (Setting Number) listed next to a G code, that setting will in some way relate to that G code.

A single asterisk (*) indicates that it's the default G code in a group.
A double asterisk (**) indicates available options.

Haas CNC Lathe: G-Codes

Code, Group, Function

G00, 01, * Rapid Positioning Motion (X, Z, U, W, B) (Setting 10,56,101)
G01, 01, Linear Interpollation Motion (X, Z, U, B, F)
G01, 01, Linear Motion with Chamfer and Cornering (X, Z, U, W, B, I, K, R, A, F)
G02, 01, CW Circular Interpolation Motion (X, Z, U, W, I, K, R, F)
G03, 01, CCW Circular Interpolation Motion (X, Z, U, W, I, K, R, F)
G04, 00, Dwell (P) (P = Seconds", "Milliseconds)
G05, 00, ** Fine Spindle Control Motion (X, Z, U, W, R, F) (Live Tooling) (Option)
G09, 00, Exact Stop, Non-Modal
G10, 00, Programmable Offset Setting (X, Z, U, W, L, P, Q, R)
G14, 00, ** Main-Spindle / Sub-Spindle Swap (Option)
G15, 00, ** Main-Spindle / Sub-Spindle Swap Cancel (Option)
G17, 02, ** Circular Motion XY Plane Selection (G02, G03) (Live Tooling) (Option)
G18, 02, * Circular Motion ZX Plane Selection (G02, G03) (Setting 56)
G19, 02, ** Circular Motion YZ Plane Selection (G02, G03) (Live Tooling) (Option)
G20, 06, * Verify Inch Coordinate Positioning (Setting 9 = Inch)
G21, 06, Verify Metric Coordinate Positioning (Setting 9 = MM)

Haas CNC Lathe: G-Codes

Code, Group, Function

G28, 00, Rapid To Machine Zero Thru Ref. Point (X, Z, U, W, B) (Fanuc)
G29, 00, Move To Location Thru G29 Ref. Point (X, Z) (Fanuc)
G31, 00, ** Feed Until Skip Function (X, Z, U, W, F) (Option)
G32, 01, Thread Cutting Path, Modal (X, Z, U, W, F)
G40, 07, * Tool Nose Compensation Cancel G41/G42 (X, Z, U, W, I, K) (Setting 56)
G41, 07, Tool Nose Compensation Left (X, Z, U, W) (Setting 43,44,58)
G42, 07, Tool Nose Compensation Right (X, Z, U, W) (Setting 56)
G50, 11, Spindle Speed Maximum RPM Limit
G51, 11, Rapid To Machine Zero, Cancel Offset (Yasnac)
G52, 00, Work Offset Positioning Coordinate (Setting 33, Yasnac)
G52, 00, Global Work Offset Coordinate System Shaft (Setting 33, Fanuc)
G53, 00, Machine Zero Positioning Coordinate, Non-Modal (X,Z,B)
G54*, 12, Work Offset Positioning Coordinate #1 (Setting 56)
G55, 12, Work Offset Positioning Coordinate #2
G56, 12, Work Offset Positioning Coordinate #3
G57, 12, Work Offset Positioning Coordinate #4
G58, 12, Work Offset Positioning Coordinate #5
G59, 12, Work Offset Positioning Coordinate #6
G61, 13, Exact Stop, Modal (X, Z)
G64, 13, * Exact Stop G61 Cancel (Setting 56)
G65, 00, ** Macro Sub-Routine Call (Option)
G70, 00, Finishing Cycle (P, Q)
G71, 00, O.D / I.D Stock Removal (P, Q, U, W, I, K, D, S, T, R1 ,F) (Setting 72,73)
G72, 00, End Face Stock Removal (P, Q, U, W, I, K, D, S, T, R1, F) (Setting 72,73)
G73, 00, Irregular Path Stock Removal Cycle (P, Q, U, W, I, K, D, S, T, F)
G74, 00, Face Groove / High Speed Peck Drill (X, Z, U, W, I, K, D, F) (Setting 22)
G75, 00, O.D / I.D Peck Grooving Cycle (X, Z, U, W, I, K, D, F) (Setting 22)
G76, 00, Thread Cutting (X, Z, U, W, I, K, A, D, F) (Setting 86,95,96,99)
G77, 00, ** Flatting Cycle (I, J, L, R, S, K) (Live Tooling) (Option)
G80, 09, * Cancel Canned Cycle (Setting 56)
G81, 09, Drill Canned Cycle (X, Z, W, R, F)
G82, 09, Spot Drill / Counterbore Canned Cycle (X, Z, W, P, R, F)
G83, 09, Peck Drill Deep Hole Cycle (X, Z, W, I, J, K, Q, P, R, F)
G84, 09, Tapping Canned Cycle (X, Z, W, R, F)
G85, 09, Bore in-Bore Out Canned Cycle (X, Z, U, W, R, L, F)
G86, 09, Bore In-Stop-Rapid Out Canned Cycle (X, Z, U, W, R, L, F)
G87, 09, Bore In-Stop-Manual Retract Canned Cycle (X, Z, U, W, R, L, F)
G88, 09, Bore In-Dwell-Manual Retract Canned Cycle (X, Z, U, W, P, R, L, F)
G89, 09, Bore In-Dwell-Bore Out Canned Cycle (X, Z, U, W, P, R, L, F)
G90, 01, O.D / I.D Turning Cycle, Model (X, Z, U, W, I, F)
G92, 01, Threading Cycle, Model (X, Z, U, W, I, F) (Setting 95,96)

Haas CNC Lathe: G-Codes

Code, Group, Function

G94, 01, End Facing Cycle, Model (X, Z, U, W, K, F) (Option)
G95, 09, ** End Face Live Tooling Rigid Tap (X, Z, U, W, K, F)
G96, 12, Constant Surface Speed, CSS On (S)
G97, 12, Constant Non-Varying Spindle Speed, Off (S) (Setting 56)
G98, 05, Feed Per Minute (F)
G99*, 05, Feed Per Revolution (F) (Setting 56)
G100, 00, Mirror Image Cancel G101
G101, 00, Mirror Image (X, Z) (Setting 45,47)
G102, 00, Programmable Output To RS-232 (X, Ż)
G103, 00, Limit Block Look-ahead (P0 – P15 = # of Blocks Look-ahead)
G105, 00, Servo Bar Command
G110, 12, Work Offset Positioning Coordinate #7
G111, 12, Work Offset Positioning Coordinate #8
G112, 00, ** Cartesian To Polar (X, Y To X,C) Transformation (Option)
G113, 00, ** Cartesian To Polar (X, Y to X,C) Transformation Cancel (Option)
G114, 12, Work Offset Positioning Coordinate #9
G115, 12, Work Offset Positioning Coordinate #10
G116, 12, Work Offset Positioning Coordinate #11
G117, 12, Work Offset Positioning Coordinate #12
G118, 12, Work Offset Positioning Coordinate #13
G119, 12, Work Offset Positioning Coordinate #14
G120, 12, Work Offset Positioning Coordinate #15
G121, 12, Work Offset Positioning Coordinate #16
G122, 12, Work Offset Positioning Coordinate #17
G123, 12, Work Offset Positioning Coordinate #18
G124, 12, Work Offset Positioning Coordinate #19
G125, 12, Work Offset Positioning Coordinate #20
G126, 12, Work Offset Positioning Coordinate #21
G127, 12, Work Offset Positioning Coordinate #22
G128, 12, Work Offset Positioning Coordinate #23
G129, 12, Work Offset Positioning Coordinate #24
G154, 12, Select Work Offset Position Coordinate, P1 - P99
G159, 00, ** Background Pickup / Part Return (Option)
G160, 00, ** APL Axis Command ON (Option)
G161, 00, ** APL Axis Command OFF (Option)
G184, 00, ** Reverse Tapping Canned Cycle (X, Z, W, R, F) (Setting 130)
G186, 00, Live Tooling Reverse Rigid Tapping
G187, 00, ** Accuracy Control For High Speed Machining (E) (Setting 85)
G194, 00, Sub-Spindle / Tapping Canned Cycle
G195, 00, ** Radial Tapping (X, F) (Live Tooling) (Option)
G196, 00, ** Radial Tapping Reverse (X, F) (Live Tooling) (Option)
G200, 00, Index On The Fly (X, Z, U, W, T)

Haas CNC Lathe: G-Codes

Note: The control automatically recognizes these G codes when your HAAS lathe is powered up:

Code, Function

G00*	Rapid Traverse
G18*	X Z Circular Plane Selection
G40*	Cutter Compensation Cancel
G54*	Work Coordinate Zero #1 (1 of 26 available)
G64*	Exact Stop Cancel
G80*	Canned Cycle Cancel
G97	Constant Surface Speed Cancel
G99*	Feed Per Revolution

Haas CNC Lathe
Letter Address Codes

B   Linear B-Axis Motion (Tailstock): (Setting 93,94,105,106,107,121,145):

The B address character is currently reserved for the tailstock. It is used to specify absolute position or motion for the tailstock along the B-axis. B-axis commands in the negative direction moves the tailstock toward the spindle, and a B-axis command in the positive direction moves it away from the spindle.

F   Feed Rate (Setting 19,77):

The F address character is used to select feed rate applied to any interpolating G codes or canned cycles. This command value is in, inches per revolution or mm per revolution. Inches per revolution (G99) is the default.

But it can be changed to units/minute with G98 traditionally, the F code was capable of only 4 fractional position accuracy; but on this control you can specify F to six place accuracy.

Code E and F are equivalent.

G   Preparatory Functions (G codes):

The G address character is used to specify the type of operation to occur in the block containing the G code. The G is followed by a two, or three-digit number between 0 and 187. Each G code defined in this control is part of a group of G codes.

G   Preparatory Functions (G codes):

The Group O codes are non-modal; that is, they specify a function applicable to this block only and do not affect other blocks. The other groups are modal and the specification of one code in the group cancels the previous code applicable from that group. A model G code applies to all subsequent blocks so those blocks do not need to re-specify the same G code. More than one G code can be placed in a block in order to specify all of the setup conditions for an operation.

I   Circular Interpolation / Canned Cycle Data:

The I address character is used to specify data used for some canned cycles and circular motions. It is either in inches with four fractional positions or mm with three fractional positions.

J   Canned Cycle Data:

The J address character is used to specify data used for some canned cycles. K   Circular Interpolation / Canned Cycle Data: The K address character is used to specify data used for some canned cycles and circular motions. It is formatted just like the I data.

Haas CNC Lathe
Letter Address Codes

L Loop Count To Repeat A Command Line:

The L address character is used to specify a repetition count for some canned cycles and auxiliary functions.

M Miscellaneous Functions M Codes:

The M address character is used to specify an M code for a block. These codes are used to control miscellaneous machine functions. Note that only one M code is allowed per block of the CNC program and all M codes are performed at the end of the block.

N Line / Block Number:

The N address character is entirely optional. The only function of a
N number is to identify and locate a certain block or line within a program.

O Program Number (Program name in parenthesis):

The O address character is used to identify a program. It is followed by a number between 0 and 99999. a program saved in memory always has a Onnnnn identification in the first block. Altering the Onnnnn in the first block causes the program to be renumbered. If you enter a program name (Name) between parenthesis in the first three lines of a program, that program name will also be seen in your list of programs. You can have up to 500 program numbers (200 programs on an older machine) in your List of Programs.

P Delay of Time / M97 Sequence Number Call / M98 Program Number Call / Live Tooling Spindle Speed:

The P address character is used as a delay of time in seconds for a dwell command, or as a P number to search for a sequence number in a local subroutine call, or as a P number to search for a sequence number in a local subroutine call, or as a P number to search for a program number in your list of programs for a subprogram call.

P is also defined with Q, and is used in canned cycles G70, G71, G72 and G73 to specify the starting block number of the part geometry defined for machining with these cycles.

Q Canned Cycle Data: The Q address charter is used in a G83 canned and is a positive number for the peck amount. Q is also defined with P, in the canned cycles G70, G71, G72 and G73 to specify the ending block number of the part geometry defined for machining with these cycles.

R Circular Interpolation / Canned Cycle Data, Setting 52:

The R address character is used in canned cycles and circular interpolation. It is usually used to define the reference plane for canned cycles.

S Spindle Speed Command, Setting 20, 144 :

The S address character is used to specify the spindle speed. The S command does not turn the spindle on or off; it only sets the desired speed. By default, S specifies RPM. When used with G96, S specifies surface feet per minute.

T Tool Selection Code, Setting 42, 87, 97:

The T number address calls a tool and an offset when initiating a tool change. Txxyy is the T command format. The first two digits (xx) specifies the turret position and is used to call up a tool that's between 1 and the number of tool turret positions on the machine. The second two digits (yy) calls up a tool geometry/were offset that are going to be used for that tool and will be a number between 1-50.

Haas CNC Lathe
Letter Address Codes

T Tool Selection Code, Setting 42, 87, 97:

The Txxyy code selects the tool and offset number that are going to be used. The T numbers differ slightly depending on if Setting 33, Coordinate System, is set for FANUC or YASNAC. Most people set it as FANUC and is what we ship the Haas machines out as from Haas. When set to FANUC the "xx" calls up a tool number (any xx leading zero numbers are omitted); the yy specifies tool geometry and were offset that are going to be used for that tool. A T101 would be recognized as tool 1 offset 1; T323 would be recognized as tool 3 offset 23. if you command T300, that would call tool 3 and cancel any active offsets. When you rapid to a location from machine home with no offsets, it'll be a minus position value in X and Z-axis or, it'll be from a part zero origin location, that'll be using a tool geometry "Txxyy" offset command.

U Incremental X-Axis Motion:

The U address character is used to specify motion for the X-axis.

It specifies an incremental position or distance along the X-axis relative to the current machine position. It is defined either in inches with four fractional positions or in mm with three fractional positions.

W Incremental Z-Axis Motion:

The W address character is used to specify motion for the Z-axis. It specifies an incremental position or distance along the Z-axis relative to the current machine position. It is defined either in inches with four fractional positions or in mm with three fractional positions.

X Absolute X-Axis Motion:

The X address character is used to specify absolute motion for the X-axis.

It specifies a position or distance along the X-axis. It is either in inches with four fractional positions or in mm with three fractional positions.

Z Absolute Z-Axis Motion:

The Z address character is used to specify absolute motion for the Z-axis.

It specifies a position or distance along the Z-axis. It is either in inches with four fractional positions or in mm with three fractional positions.

Note: For all motion commands, if the decimal point is omitted, the value is in units of .0001 inch or .001 mm.

Haas CNC Lathe: M-Codes

All M codes are activated or cause an action to occur after everything else on a block has been completed. Only one M code is allowed per block in a program.

If there is a (Setting Number) listed next to an M code. That setting will in some way relate to the M code.

A single asterisk (*) indicates that it's the default G code in a group.
A double asterisk (**) indicates options available.

Code, Function

M00, Program Stop (Setting 42, 101)
M01, Optional Program Stop (Setting 17)
M02, Program End
M03, Spindle On Forward (S) (Setting 144)
M04, Spindle On Reverse (S) (Setting 144)
M05, Spindle Stop
M08, Coolant On (Setting 32)
M09, Coolant Off
M10, Chuck Clamp (Setting 92)
M11, Chuck Unclamp (Setting 92)
M12,** Auto Air Jet ON (P) (Option)
M13,** Auto Air Jet OFF (Option)
M14,** Main Spindle Clamp (Option)
M15,** Main Spindle Unclamp (Option)
M17, Rotate Turret Forward (T) (Setting 97)
M18, Rotate Turret Reverse (T) (Setting 97)
M19,** Orient Spindle (P, R) (Option)
M21,** Tailstock Advance (Setting 93,94,106,107,121,145) (Option)
M22,** Tailstock Retract (Setting 105) (Option)
M23, Angle Out Of Thread ON (Setting 95,96)
M24, Angle Out Of Thread OFF
M30, Program End And Reset (Setting 2,39,56,83)
M31,** Chip Auger Forward (Setting 114,115)
M33, Chip Auger Stop
M36,** Parts Catcher ON (Option)
M37,** Parts Catcher OFF (Option)
M41, Spindle Low Gear Override
M42, Spindle High Gear Override
M43, Turret Unlock (For Service Use Only)
M44, Turret Lock (For Service Use Only)
M51, Optional User M-Code Set
M52, Optional User M-Code Set

Haas CNC Lathe: M-Codes

Code, Function

M53, Optional User M-Code Set
M54, Optional User M-Code Set
M55, Optional User M-Code Set
M56, Optional User M-Code Set
M57, Optional User M-Code Set
M58, Optional User M-Code Set
M59, Output Relay Set (N)
M61, Optional User M-Code Clear
M62, Optional User M-Code Clear
M63, Optional User M-Code Clear
M64, Optional User M-Code Clear
M65, Optional User M-Code Clear
M66, Optional User M-Code Clear
M67, Optional User M-Code Clear
M68, Optional User M-Code Clear
M69, Output Relay Clear (N)
M76, Program Displays Inactive
M77, Program Displays Active
M78, Alarm if Skip Signal Found
M79, Alarm if Skip Signal Not Found
M85**, Automatic Door Open (Setting 51,131) (Option)
M86**, Automatic Door Close (Setting 51,131) (Option)
M88**, High Pressure Coolant ON (Setting 32) (Option)
M89**, High Pressure Coolant OFF (Option)
M93**, Axis Position Capture Start (P, Q) (Option)
M94**, Axis Position Capture Stop (Option)
M95, Sleep Mode (HH:MM)
M96, Jump if No Signal (P, Q)
M97, Local Sub-Program Call (P, L)
M98, Sub-Program Call (P, L)
M99, Sub-Program Return Or Loop (P) (Setting 118)
M109, Interactive User Input (P) (Option)
M110**, Sub-Spindle Chuck Clamp (Setting 122) (Option)
M111**, Sub-Spindle Chuck Unclamp (Setting 122) (Option)
M119**, Sub-Spindle Orient (P, R) (Option)
M121, Optional User M-Code Interface With M-Fin Signal
M122, Optional User M-Code Interface With M-Fin Signal
M123, Optional User M-Code Interface With M-Fin Signal
M124, Optional User M-Code Interface With M-Fin Signal
M125, Optional User M-Code Interface With, M-Fin Signal
M126, Optional User M-Code Interface With, M-Fin Signal

Haas CNC Lathe: M-Codes

Code, Function

M127, Optional User M-Code Interface With M-Fin Signal
M128, Optional User M-Code Interface With M-Fin Signal
M133,* Live Tool Drive Forward (P) (Option)
M134,* Live Tool Drive Reverse (P) (Option)
M135,* Live Tool Drive Stop (Option)
M143,* Sub-Spindle Forward (P) (Option)
M144,* Sub-Spindle Reverse (P) (Option)
M145,* Sub-Spindle Stop (Option)
M154,* C Axis Engage (Setting 102) (Option)
M155,* C Axis Disengage (Option)
M164, Rotate APL Grippers To "N" Position (P, N) (Option)
M165, Open APL Grippers 1 (Raw Material) (Option)
M166, Close APL Grippers 1 (Raw Material) (Option)
M167, Open APL Grippers 2 (Finished Material) (Option)
M168, Close APL Grippers 2 (Finished Material) (Option)

Haas CNC SL-10 Lathe

Haas CNC SL-20 Lathe

Haas CNC Lathe
Program Structure

A CNC part program consists of one or more blocks of commands. When viewing the program, a block is the same as a line of text. Blocks shown on the computer screen are always terminated by the " ; " (semicolon) symbol which is called an End Of Block (EOB). Blocks are made up of alphabetical address codes, which are always a letter followed by a number. For instance, the specification to move the X-axis would be a number preceded by the X symbol, such as X2.750 the number can also have a negative value X-2.75.

Programs must begin and end with a percent (%) sign. After the first percent (%) sign with nothing else on the that line, the next line in a program must have a program number beginning with the letter O (Not Zero) and then up to five numbers (o12345) that defines that program. Those program numbers are used to identify and select a main program to be run, or as a subprogram called up by the main program. The % sign will not be seen on the control. But they must be in the program when you load a program into the Haas control. And they will be seen when you download a program from the machine. The % signs are automatically entered in for you, if you enter a program in on the Haas control.

A program may also contain a " / " symbol. The " / " symbol, sometimes called a slash or forward-slash, is used to define an optional block. If a block contains this symbol, any information that follows the slash in a program block, will be ignored when the Block Delete button is selected when running a program.

On the following page is a sample program, as it would appear on the control screen. The words in parentheses following the blocks are comments and will not be read by the Haas control. Comments must always be in parentheses so they will not be read by the Haas control.

This sample program will rough and finish turn and face for a part with two diameters along with drilling and tapping for a 3/8-16 x 1.0 deep threaded hole, one end. The program with comment statements would appear as follows. Programs are normally written in notepad and notepad will automatically add the (;) (semicolon) at end of block.

Haas CNC Lathe
Program Structure

%
o00018
N1
G28
T101
G50 S2600
G97 S414 M03
G54 G00 X3.6 Z0.1 M08
G96 S390
G00 Z0.005
G01 X-0.063 F0.005
G00 X3.6 Z0.1
G71 P10 Q20 U0.01 W0.005 D0.1 F0.01
N10 G42 G00 X0.82
G01 Z0. F0.004
X0.9
G03 X1. Z-0.05 R0.05
/G01 Z-1.75
/X1.75
/G03 X2.25 Z-2. R.25
G01 Z-3.25 F0.004
X2.94
X3. Z-3.28
Z-4.1
N20 G40 X3.6
G97 S414 M09
G28
M01
(Continued in next Column)

(Continued from previous Column)
N2
G28
T202
G50 S2600
G97 S1354 M03
G54 G00 X1.1 Z0.1 M08
G96 S390
G00 Z0.
G01 X-0.032 F.003
G00 X3.6 Z0.1
G70 P10 Q20
G97 S414
G00 Z1.0 M09
G28
M01
N3
G28
T303
G97 S1986 M03
G54 G00 X0. Z1. M08
G83 Z-2.5 Q0.3 R0.1 F0.005
G80 G00 Z0.1 M09
G28
T100
M30
%

Haas CNC Lathe
G00 Rapid Position Command

G00 Rapid Positioning motion
X = Absolute X-axis positioning command
Z = Absolute Z-axis positioning command
U = Incremental X-axis positioning command
W = Incremental Z-axis positioning command
B = Absolute tailstock command

You need to be careful and avoid obstructions with this type of rapid move. The tool will first move from the current position in a straight line along a 45-degree angle to an intermediate location, until one of these axes has completed its move. Then the machine will move parallel to the X or Y-axis to complete the rapid move to the final location. These rapid moves may be in absolute or incremental coordinate command values, which will change how those values are interpreted. The "U" letter address relates to X-axis incremental moves and the "w" letter address relates to Z-axis incremental moves.

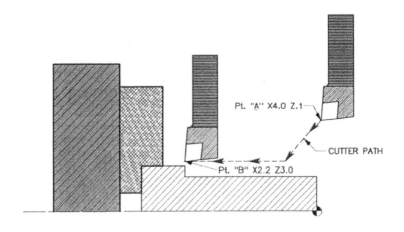

Move from point "A" to point "B" can be G00 X2.2 Z-3.0
or G00 U-1.8 W-3.1
or G00 X2.2 W-3.1
or G00 U-1.8 Z-3.0

Haas CNC Lathe
G01 Linear Interpolation Command

G01 Linear Interpolation Motion
X = Absolute X-axis motion command
Z = Absolute Z-axis motion command
U = Incremental X-axis motion command
W = Incremental Z-axis motion command
F = Feed rate in inches (or mm) per minute

Motion can occur in 1 or 2 axes. Both axes will start and finish motion at the same time to move the tool along a straight-line path parallel to an axis or at a slope (angled) line. The speeds of all axes are controlled so that the feed-rate specified is achieved along the actual path. The F (feed-rate) command is modal and may be specified in a previous block. These moves may be made in absolute or incremental coordinate command values, which change how those values are interpreted. When feeding to a location using G01, it should be using previously defined tool geometry and tool offset command (Txxyy).

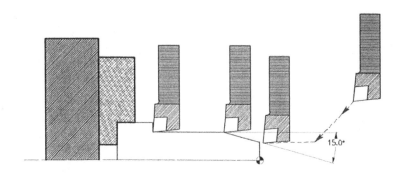

G00 X0.9106 Z0.1
(Absolute)
G00 X1.5 Z-1. F0.00G
Z-3.
X2.

G00 X0.9106 Z0.1
(Incremental)
G01 U.5894 W-1.1 F0.006
W-1.9
U0.5

G00 X0.9106 Z0.1
(Absolute & Incremental)
G01 X1.5 W-1.1 F0.006
Z-3.
U0.5

Haas CNC Lathe
G01 Linear Interpolation

Linear interpolation to feed in a straight-line motion from point to point for machining diameters, faces and corner chamfering in absolute or incremental. With G01 we can move the edge of a cutter along the part profile, by a series of absolute X and Y-axis movements, or incremental U (X-axis) and/or incremental W (Z-axis) movements.

The part can be programmed with both X and Z absolute movements and/or incremental U and W movements. You can define the actual points around part using both incremental and absolute.

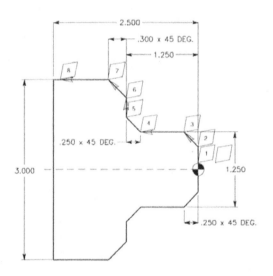

Absolute
Programming

N . . G00 X0.25 Z0.1
N11 G01 Z0. F0.00
N12 X.750
N13 X1.250 Z-.250
N14 Z-1.000
N15 X1.750 Z-1.250
N16 X2.400
N17 X3.000 Z-1.550
N18 Z-2.375
N19 G00 X3.01 Z0.1

Absolute & Incremental
Programming

N . . G00 X0.25 Z0.1
N21 G01 Z0. F0.006
N22 X.750
N23 U.500 W-.250
N24 W-.750
N25 U.500 W-.250
N26 U.650
N27 U.600 W-.300
N28 W-.825
N29 G00 U0.01 Z0.1

Haas CNC Lathe
G01 & G02 Circular Interpolation

G01 & G02 Using I & K

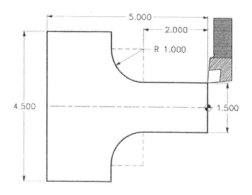

N11 G01 Z-2.00 F0.01
N12 G02 X3.500 Z-3.000 I1.000 K0
N13 G01 X4.500

G01 & G02 Using I & K

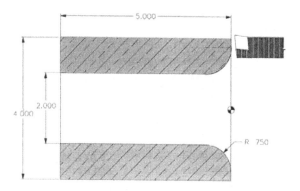

N21 G01 Z0. F0.01
N22 X3.000
N23 G02 X2.000 Z-.750 I0 K-.750
N24 G01 Z-5.000

Haas CNC Lathe
G01 & G03 Circular Interpolation

G01 & G03 Using I & K

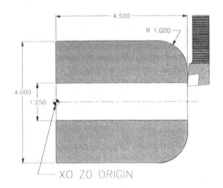

N31 G01 X2.000 F0.01
N32 G03 X4.000 Z3.500 I0 K-1.000
N33 G00 Z3.500

G01 & G03 Using I & K

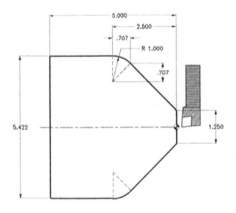

N41 G01 F0.01
N42 X1.250
N43 X3.908 Z-1.973
N44 G02 X4.836 Z-2.500 I-.707 K-.707
N45 G01 Z-5.000

Haas CNC Lathe
G02 & G03 Circular Interpolation

Five pieces of information are required for executing a circular interpolation command: Plane selection, arc start position coordinates, rotation direction, arc end position coordinates, and arc center coordinates or arc radius.

G02 & G03 Circular Interpolation Commands

X	Absolute circular end point X-axis motion command
Z	Absolute circular end point Z-axis motion command
U	Incremental circular end point X-axis motion command
W	Incremental circular end point Z-axis motion command
I	X-axis Incremental distance from the start point to arc center (If R is not used)
K	Z-axis Incremental distance from the start point to arc center (If R is not used)
R	Radius of the arc (If I and K are not used)
F	Feed rate in inches (or mm) per minute

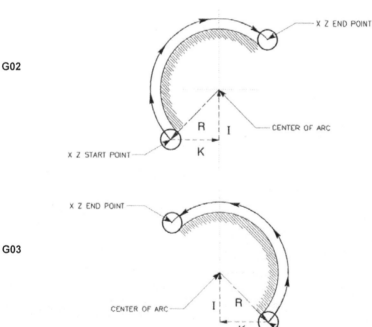

G02

G03

Haas CNC Lathe
G02 & G03 Circular Interpolation Using I & K or R

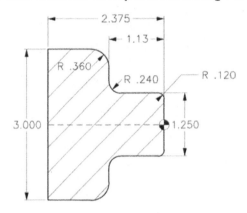

G02 & G03 Using I & K
N11 G01 Z0. F0.005
N12 X1.01
N13 G03 X1.250 Z-0.120 I0. K-0.120
N14 G01 Z-0.89
N15 G02 X1.75 Z-1.13 I0.240 K0.
N16 G01 X2.28
N17 G03 X3. X-1.49 I0. K-.360
N18 G01 Z-2.375

G02 & G03 Using R
N21 G01 Z0. F0.005
N22 X1.01
N23 G03 X1.25 Z-0.12 R0.120
N24 G01 Z-0.89
N25 G02 X1.73 Z-1.13 R.240
N26 G01 X2.28
N27 G03 X3. Z-1.49 R.360
N28 G01 Z-2.375

Manual Program and Compensation of Part Radius Tool Has .031 TNR,

G02 & G03 Using I & K
%
o00052
N101 G53 G00 X0. Z0. T0
N102 T101 (O.D. Tool)
N103 G97 S1450 M03
N104 G54 G00 X0.91 Z0.1
N105 G01 Z0. F0.01
N106 X0.948 F0.006
N107 G03 X1.25 Z-0.151 I0. K-0.151
N108 G01 Z-0.921
N109 G02 X1.668 Z-1.13 I.209 K0.
N110 G01 X2,218
N111 G03 X3. Z-1.521 I0. K-0.391
N112 G01 Z-2.375
N113 G00 U0.01 Z1.0
N114 G54 G00 X0. Z0. T0
N115 M30
%

G02 & G03 Using R
%
o00052
N201 G53 G00 X0. Z0. T0
N202 T101 (O.D. Tool)
N203 G97 S1450 M03
N204 G54 G00 X0.91 Z0.1
N205 G01 Z0. F0.01
N206 X0.984 F0.006
N207 G03 X1.25 Z-0.151 R0.151
N208 G01 Z-0.921
N209 G02 X1.668 Z-1.13 R0.209
N210 G01 X2.218
N211 G03 X3. Z-1.521 R0.391
N212 G01 Z-2.375
N213 G00 U0.01 Z1.0
N214 G54 G00 X0. Z0. T0
N215 M30
%

Haas CNC Lathe
G02 & G03 Circular Interpolation Motion

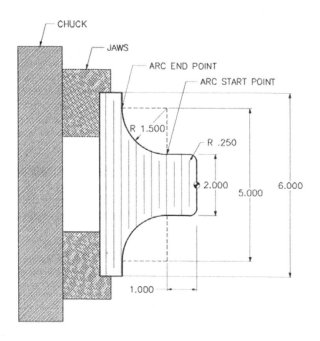

This program example starts machining with the tool at the beginning of the .250 radius to machine in a counter- clockwise direction. Then moves cutter to the start of 1.500 radius and around in a clockwise direction from start-point to end-point. For each radius, the programming code would look like this:

```
N4 . . . . .
N5 G00 X1.4 Z0.1
N6 G01 Z0. F0.008
N7 X1.5
N8 G03 X2. Z-0.25 R0.25 (or G03 X2. Z-0.25 I0. K-0.25)
N9 G01 Z-1.
N10 G02 X5. Z-2.5 R1.5 (or G02 X5. Z-2.5 I1.5 K0.)
N11 G01 X6.0
N12 G00 Z0.1
N . . . . .
```

Haas CNC Lathe
G01 Corner Rounding Using R

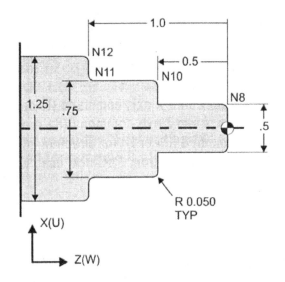

```
%
o00042 (Liner G01 with Radius using R)
N1 G53 G00 X0. Z0. T0
N2 T101 (O.D. Tool x .031 TNR
N3 G50 S3000
N4 G97 S3000 M03
N5 G54 G00 X0.3 Z0.1 M08
N6 G96 S390
N7 G42 G01 Z0. F0.01
N8 G01 X0.5 R-0.05
N9 G01 Z-0.5
N10 G01 X0.75 R-0.05
N11 G01 Z-1.0 R0.05
N12 G01 X1.25 R-0.05
N13 G01 Z-1.5
N14 G40 G00 U0.01 Z0.1 M09
N15 G53 G00 X0. Z0. T0
N16 M30
%
```

Haas CNC Lathe
G01 Chamfering 45-degree Angles

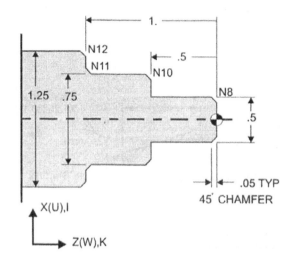

```
%
o00043 (Liner G01 with 45-degree Chamfer using I or K)
N1 G53 G00 X0. Z0. T0
N2 T101 (O.D. Tool x .031 TNR
N3 G50 S3000
N4 G97 S3000 M03
N5 G54 G00 X0.3 Z0.1 M08
N6 G96 S390
N7 G42 G01 Z0. F0.01
N8 G01 X0.5 K-0.05
N9 G01 Z-0.5
N10 G01 X0.75 K-0.05
N11 G01 Z-1.0 I0.05
N12 G01 X1.25 K-0.05
N13 G01 Z-1.5
N14 G40 G00 U0.01 Z0.1 M09
N15 G53 G00 X0. Z0. T0
N16 M30
%
```

Haas CNC Lathe
G01 Auto Chamfering

1) The linear G01 block must be single X(U) or Z(W) move that is perpendicular to the previous move with an A to do a specific angle.

2) When using A for an angle, do not use I, K or R.

3) This angle (A) command is not supported in roughing passes of G71 or G72 canned cycles, though the last pass of G71 or G72 will execute the G01 "A" angle.

4) A G70 or G73 will support this type of chamfer com.

5) You can use a minus value to define an angle clockwise from the three o-clock position.

6) be sure to enter a decimal point for angles.

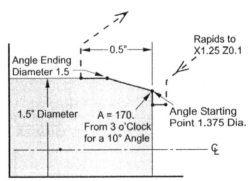

```
%
o00044
N1 G53 G00 X0. Z0. T0
N2 T101 (O.D. Tool x .031 TNR)
N3 G50 S2800
N4 G97 S1490 M03
N5 G00 X1.25 Z0.1 M08
N6 G96 S390
N7 G42 G01 Z0. F0.01
N8 G01 X1.375 (Start point)
N9 G01 X1.5 A170. (10-degree angle using A)
N10 Z-0.5
N11 G40 G00 U0.01 Z0.1 M09
N12 G53 G00 X0. Z0. T0
N13 M30
%
```

Haas CNC Lathe
Chamfering & Rounding
Interpolation with G01, G02 and G03

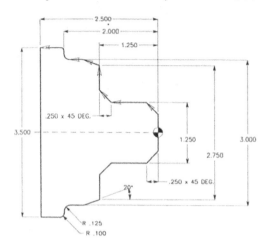

```
%
o00045
N11 G53 G00 X0. Z0. T0
N12 T101 (O.D. Tool x .031 TND)
N13 G50 S2600
N14 G97 S2057 M03
N15 G54 G00 X0.65 Z0.1
N16 G96 S350
N17 G01 G42 Z0. F0.006
N18 X.75
N19 X1.25 Z-0.25
N20 Z-1.0
N21 X1.75 Z-1.25
N22 X2.75
N23 X3. Z-1.5934 (Calculated Z move)
N24 Z-1.875
N25 G02 X3.25 Z-2.0 R0.125
N26 G01 X3.3
N27 G03 X3.5 Z-2.1 R0.1
N28 G01 Z-2.375
N29 G00 G40 U0.01 Z0.1 M09
N30 G53 G00 X0. Z0. T0
N31 M30
%
```

Haas CNC Lathe
Chamfering & Rounding
G01 Interpolation using I, K, R, and A

This G01 Code using I, K, R, or A, to do an arc or angle, is not recognized in the roughing passes of the canned cycles G71 or G72.

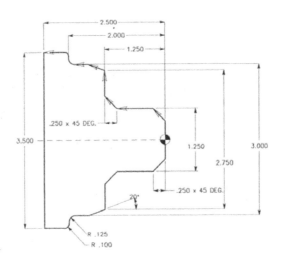

```
%
o00050
N31 G53 G00 X0. Z0. T0
N32 T101
N33 G50 S2600
N34 S97 S2057 M03
N35 G54 G00 X0.65 Z0.1
N36 G96 S350
N37 G42 G___ Z0. F0.006
N38 X___ K___ (K+ or -)
N39 Z___ I___ (K+ or -)
N40 X___ (To X start point for A)
N41 X___ A___ (K+ or -)
N42 Z___ R___ (K+ or -)
N43 X___ R___ (K+ or -)
N44 Z___
N45 G40 G00 U0.01 Z0.1 M09
N46 G53 G00 X0. Z0. T0
N47 M30
%
```

Haas CNC Lathe
G41 & G42 TNR Compensation
Manually Calculating TNR Compensation

When you program a straight line in either X or Z the tool tip touches the part at the same point where you touched your original tool offsets in X and Z. however, when you program a radius, the theoretical tool tip does not touch the part radius. Where the tip actually touches the part is dependent upon the radius of the tool and the point around the radius being cut.

90-degree corner radius & .031 TNR in your program. Manually calculate the compensated path by adding .031 to the radius to be machined. You will also need to recalculate the start point and end point of this larger programmed radius.

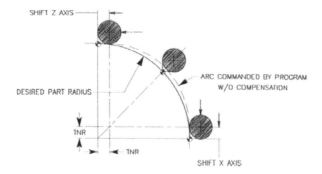

90-degree fillet radius & .031 TNR in your program. Manually calculate the compensated path by subtracting .031 from the radius to be machined. You will also need to recalculate the start point and end point of this smaller programmed radius.

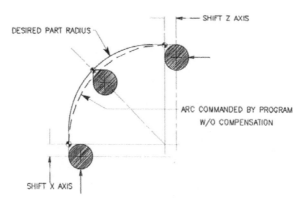

Haas CNC Lathe
G41 & G42 TNR Compensation
Tool Nose Radius Calculation Diagram

Machining a chamfer, when cutter compensation is NOT used on the control, requires that calculations must be made for the tool tip geometry for the programmed moves on the part angles.

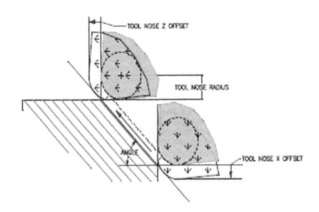

In the lathe programming and operation manual there is a section for manually calculating tool nose radius compensation with various charts and illustrations for different angles for 1/32 and 1/64 insert radiuses.

Angle	X Offset .015 TNR	Z Offset .015 TNR	X Offset .031 TNR	Z Offset .031 TNR
15°	.0072	.0136	.0146	.0271
30°	.0132	.0114	.0264	.0229
45°	.0184	.0092	.0366	.0183
60°	.0228	.0066	.0458	.0132

See Cutter Path Calculations, In Data Section.

Haas CNC Lathe
G41 & G42 TNR Compensation
Manually Adding TNR Compensation

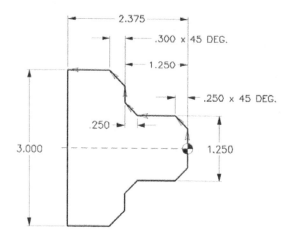

Program is manually adding in compensation for a finish pass using a tool with a 0.031 TNR.

```
%
o00055 (Linear interpolation without cutter comp.)
N1 G53 G00 X0. Z0. T0
N2 T101 (O.D. Turning tool .031 TNR)
N3 G97 S1450 M03
N4 G54 X00 X0.85 Z0.1 M08
N5 Z0.
N6 G01 X-.062 F0.01 (Face down end of part)
N7 G00 X.7134 Z0.02
N8 G01 Z0. F0.006 (Feed into face)
N9 X1.250 Z-.2683
N10 Z-1.0183
N11 X1.7134 Z-1.250
N12 X2.3634
N13 X3.0 Z-1.5683
N14 Z-2.375
N15 G00 U0.01 Z1.0 M09
N16 G53 G00 X0. Z0. T0
N17 M30
%
```

Haas CNC Lathe
G41 & G42 TNR Compensation
Seven Steps For Using Tool Nose Compensation

1) Approach moves:
Program an approach move for each tool path that needs tool nose compensation and determine if G41 or G42 is to be used.

2) Departure Moves:
Be sure there is a departure move for each compensation tool path by using a G40 command to cancel tool nose compensation.

3) Tool Geometry Offsets And Tool Wear Offsets:
When setting up the tools for a part program, zero any Tool Geometry, Tool Wear, and Work Zero offsets that remain from an earlier job. Then you must touch off and enter the tool geometry (distance from machine home to par zero) length offsets for each tool being used.

4) Tool Nose Radius Geometry:
Select a standard insert (with a defined radius) that will be used for each tool that is using tool nose compensation. Enter the tool nose radius of each compensated tool in the TOOL GEOMETRY offset display under RADIUS.

5) Tool Tip Direction:
Input the tool tip direction, in the TOOL GEOMETRY offset display under TIP, for each tool that is using tool nose compensation, G41 or G42.

6) Test Run Compensation Geometry:
Run the program in graphics mode and correct any tool nose compensation geometry problems that may occur. A problem can be detected in two ways: either an alarm will be generated indicating compensation interference, or you will see the incorrect geometry generated and seen in graphics mode.

7) Run And Inspect First Part:
Carefully cycle through the program to machine part. After running part, check and adjust were offsets for the part to bring it within size.

Haas CNC Lathe
G41 & G42 TNR Compensation
Tool Tip Orientation

Tool tip orientation T-number is only effective in G41/G42 mode and must be set at the control before the program is processed.

Tip Code	Tool Tip Orientation	Tip Code
Tip 0		Tip 5
Tip 1		Tip 6
Tip 2		Tip 7
Tip 3		Tip 8
Tip 4		Tip 9

0) Zero (0) indicates no specified direction.
1) Tool Tip off in the X+, Z+ direction.
2) Tool Tip off in the X+, Z- direction.
3) Tool Tip off in the X-, Z- direction.
4) Tool Tip off in the X-, Z+ direction.
5) Tool Edge Z+ direction.
6) Tool Edge X+ direction.
7) Tool Edge Z- direction.
8) Tool Edge X- direction.
9) Same as Tip 0.

Haas CNC Lathe
Machine Cycles For Turning, Facing & Grooving

A Machine cycle is used to simplify the programming of a part. Machine cycles are defined for the most common machining operations. They can be divided into two types. There are machining cycles for turning and grooving. And there are canned cycles for drilling, tapping, and boring. These cycles can be either single block cycles or model cycles.

A model cycle remains in effect after they are defined and are executed for each positioning of an axes. Once a cycle is defined, that operation is performed at every X-Z position subsequently listed in a block.

The machining cycles G70, G71, G72, and G73 cannot de in DNC while running a program, since these G-codes require the control to look ahead.

These G-codes, (G71, G72, and G73) do not force feed moves within the PQ block. To prevent rapid motion from occurring when feed moves are intended, a G01 should be defined, when needed, near the beginning of the P block. The feed rate that it'll do throughout the cycle is the one listed on the canned cycle line. And the G70 will perform the feed rates that are listed on the lines between the PQ blocks.

Canned cycles G70, G71, G72, and G73 should not be executed after a tool nose compensation command (G41 or G42). If tool nose compensation is desired, it should be defined in the P block called up by the canned cycle.

The Following is a list of the canned cycles that can be used for turning and grooving on Haas Lathe Controls.

Code, Group, Description
G70, 00, Finishing Cycle
G71, 00, O.D. / I.D. Stock Removal Cycle
G72, 00, End Face Stock Removal Cycle
G73, 00, Irregular Path Stock Removal Cycle
G74, 00, End Face Grooving Cycle, or Peck Drilling
G75, 00, O.D. / I.D. Grooving Cycle
G76, 00, O.D. / I.D. Thread Cutting Cycle, Multiple Pass
G90, 01, Turning Cycle, Modal
G92, 01, Thread Cutting Cycle, Model
G94, 01, End Face Cutting Cycle, Model

Haas CNC Lathe
G71 O.D. / I.D. Stock Removal Cycle

P Starting block number of part path to machine.
Q Ending block number of part path to machine.
U Finish stock remaining with direction (+ or -) X-axis diameter value. (optional)
W Finish stock remaining with direction (+or-) Z-axis value. (optional)
I Last pass amount with direction (+or-) X-axis radius value. (optional)
K Last pass amount with direction (+or-) Z-axis value. (optional)
D Depth of cut stock removal each pass, positive radius value. (setting 72) (optional)
F Roughing passes feed rate throughout this cycle
R1 Yasnac type II roughing, only if setting 33 is on Yasnac.
S Spindle speed in this cycle
T Tool and offset in this cycle

Note: S and T are rarely defined in a G71 line

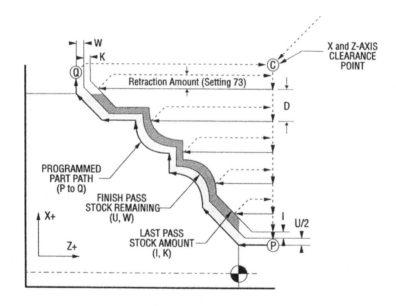

Haas CNC Lathe
G71 Type-1 Roughing I.D. & G70 Finishing I.D.

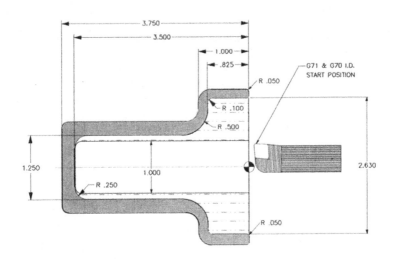

```
%
o00088
N1 G53 G00 X0. Z0. T0
N2 T404
N3 G50 S3000
N4 G97 S1780 M03
N5 G54 G00 X0.9 Z0.1 M08
N6 G96 S420
N7 G71 P8 Q18 U-0.01 W0.005 D0.12 F0.012
N8 G41 G00 X2.83
N9 G01 Z0. F0.02
N10 X2.73 F0.005
N11 G02 X2.63 Z-.05 R0.05
N12 G01 Z-.725
N13 G03 X2.43 Z-.825 R.1
N14 G01 X2.25
N15 G02 X1.25 Z-1.325 R0.5
N16 G01 Z-3.25
N17 G03 X.75 Z-3.5 R0.25
N18 G01 G40 X0.7
N19 G70 P8 Q18
N20 G97 S1780 M09
N21 G53 G00 X0. Z0. T0
N22 M30
%
```

Haas CNC Lathe
G72 End Face Stock Removal Cycle

- **P** Starting block number of part path to machine.
- **Q** Ending block number of part path to machine.
- **U** Finish stock remaining with direction (+or-) X-axis diameter value. (optional)
- **W** Finish stock remaining with direction (+or-) Z-axis value. (optional)
- **I** Last pass amount with direction (+or-) X-axis radius value. (optional)
- **K** Last pass amount with direction (+or-) Z-axis value. (optional)
- **D** Depth of cut stock removal each pass, positive radius value. (setting 72) (optional)
- **F** Roughing passes feed rate throughout this cycle
- **R1** Yasnac type II roughing, only if setting 33 is on Yasnac.
- **S** Spindle speed in this cycle
- **T** Tool and offset in this cycle

Note: S and T are rarely defined in a G72 line.

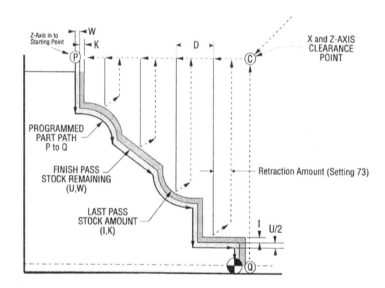

Haas CNC Lathe
G72 Type-1 Rough & G70 Finish Facing

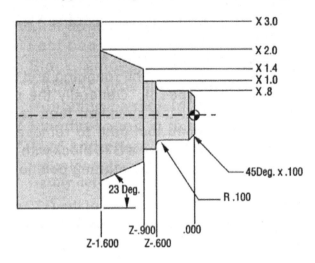

```
%
o00096
N1 (G72 Roughing face)
N2 G53 G00 X0. Z0. T0
(Sending home for tool change)
N3 T101 (O.D. Tool x .031 TNR)
N4 G50 S3000
N5 G97 S450 M03
N6 G54 G00 X3.1 Z0.1 M08
(Rapid to start point above stock)
N7 G96 S370
N8 G72 P9 Q18 U0.01 W0.01 D0.06 F0.012
(G72 Rough facing cycle with TNC)
N9 G41 G00 Z-1.6
(Starting sequence No. defined by)
 (P8 in G72 and G70)
N10 G01 X2. F0.008
N11 X1.4 Z-0.9
N12 X1.
N13 Z-0.6
N14 G03 X0.8 Z-0.5 R0.1
N15 G01 Z-0.1

(Program continued next column)

N16 X0.6 Z0.
N17 X-0.062
N18 G40 G00 Z0.1
(End of path geometry defined)
(with P17 in G72 and G70)
N19 G97 S450 M09
N20 G53 G00 X0. Z0. T0
N21 M01
N22 (G70 Finishing face)
N23 G53 G00 X0. Z0. T0
(Sending home for tool change)
N24 T202 (O.D. Tool X .015 TNR)
N25 G50 S3000
N26 G97 S450 M03
N27 G54 G00 X3.1 Z0.1 M08
(Rapid to start point)
N28 G96 S420
N29 G70 P9 Q18
(Finish face with finish cycle)
N30 G97 S450 M09
N31 G53 G00 X0. Z0. T0
(Sending home for tool change)
N32 M30
%
```

G72 is more efficient to use instead of G71 if the roughing cuts in the X-axis are longer than the roughing cuts in the Z-axis.

Haas CNC Lathe
G72 Type-1 Roughing & G70 Finishing Face

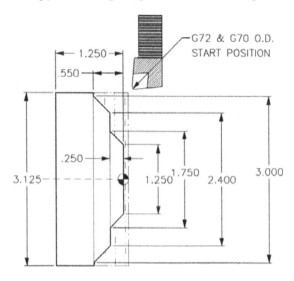

```
%
o00080
N1 (G72 Rough face)
G53 G00 X-3. Z-4. T0
T101 (O.D. Tool x .031 TNR)
G50 S2800
G97 S500 M03
G54 G00 X____ Z____ M____
G____ S____
Z____
G____ P____ Q____ U____ W____ D____ F____
N10 G____ G____ Z____
G____ X____ F.006
X____ Z____
X____
X____ Z____
X____
N20 G____ G____ Z____
```

(Continued next column)

```
G____ S____ M____
G53 G00 X-3. Z-4. T0
M____
N2 (G72 Rough face)
G53 G00 X-3. Z-4. T0
T____
G____ S
G____ S____ M____
G54 G00 X____ Z____ M____
G____ S____
Z____
G____ P____ Q____
G____ S____ M____
G53 G00 X-3. Z-4. T0
M30
%
```

Haas CNC Lathe
G73 Irregular Path Stock Removal Cycle

P Starting block number of part path to machine.
Q Ending block number of part path to machine.
U Finish stock remaining with direction (+or-) X-axis diameter value. (optional)
W Finish stock remaining with direction (+or-) Z-axis value. (optional)
I Last pass amount with direction (+or-) X-axis radius value. (optional)
K Last pass amount with direction (+or-) Z-axis value. (optional)
D Depth of cut stock removal each pass, positive radius value (setting 72) (optional)
F Roughing passes feed rate throughout this cycle
S Spindle speed in this cycle
T Tool and offset in this cycle

Note: S and T are rarely defined in a G73 line.

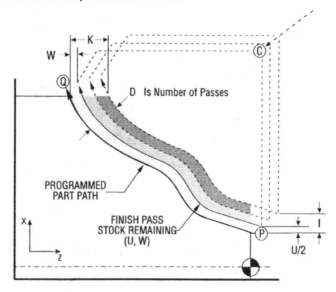

Haas CNC Lathe
G74 High Speed Peck Drill Cycle

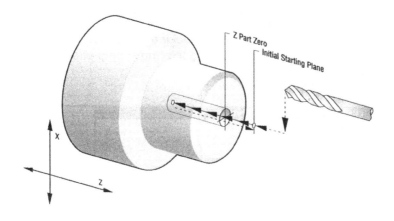

%
o00107
(G74 High speed peck drilling cycle)
(Drill .500" diameter hole to a .525" depth)
N1 G28
N2 T404 (1/2" diameter drill)
N3 G97 S2445 M03
N4 G54 G00 X0. Z0.1 M08 (Rapid to a start point)
N5 G74 Z-.525 K0.01 F0.006
(Drills to Z-.525 depth, pecking every .1" to pull back)
(after each peck the amount in Setting 22)
N6 M09
N7 G28
N8 M30
%

Setting 22 (Can cycle delta Z) – As the drill pecks deeper into the part, with each peck value of K, it pulls back a constant specified distance above the bottom of the hole created by the previous peck to break the chip. That specified distance it pulls back is defined in Setting 22.

Haas CNC Lathe
G74 End Face Grooving

- **X** X-axis absolute location to the furthest peck, diameter value. (optional)
- **Z** Z-axis absolute pecking depth.
- **U** X-axis incremental distance and direction (+ or -) to furthest peck, diameter value. (optional)
- **W** Z-axis incremental pecking depth. (optional)
- **I** X-axis shift increment between pecking cycles, positive radius value. (optional)
- **K** Z-axis pecking depth increment. (optional)
- **D** Tool shift amount when returning to clearance plane. (optional) (see note)
- **F** Feed rate.

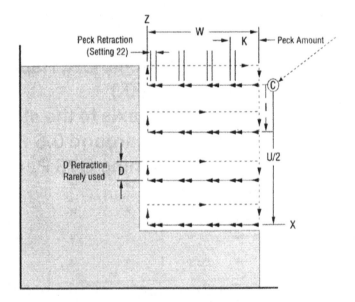

Note:
The D code command is rarely used and should only be used if the wall on the outside of groove does not exist like the diagram above shows. The D code can be used in grooving and turning to provide a tool clearance shift, in the X-axis, before returning in the Z-axis to the "C" clearance point. But, if both sides to the groove exist during the shift, then the groove tool would break. So you wouldn't want to use the D command.

Haas CNC Lathe
G74 Single Pass Face Grooving Cycle

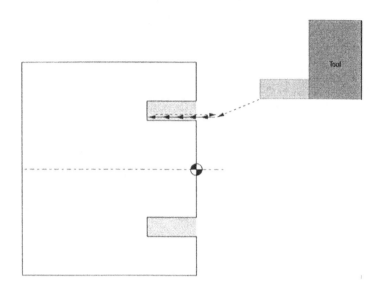

```
%
o00105
(G74 Single pass end face grooving cycle)
(Machine a .25" wide groove with a .25" groove tool)
N1 G28
N2 T404 (.25" End face groove tool)
N3 G97 S1150
N4 G54 G00 X1.5 Z0.1 M08 (Rapid to a start point)
N5 G74 Z-0.375 K0.1 F0.005
(Feed to a Z-.375 depth with a .1 peck)
N6 M09
N7 G28
N8 M30
%
```

Setting 22 (Can cycle delta Z) – As the groove tool pecks deeper into the part, with each peck value of K, it pulls back a constant specified distance above the bottom of the groove created by the previous peck to break the chip. That specified distance it pulls back is defined in Setting 22.

Haas CNC Lathe
G74 Multiple Pass Face Grooving Cycle

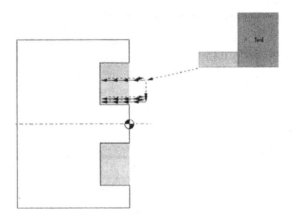

%
o00106
(G74 Multiple pass end face grooving cycle)
(Machine a .500" wide groove with a .25" groove tool)
N1 G28
N2 T404 (.25" End face groove tool)
N3 G97 S1150
N4 G54 G00 X1.75 Z0.1 M08 (Rapid to a start point)
N5 G74 X1.25 Z-0.375 I0.2 K0.05 F0.005
(G74 Multiple pass groove .05 peck Z-.375 depth)
N6 M09
N7 G28
N8 M30
%

Setting 22 (Can cycle delta Z) – As the groove tool pecks deeper into the part, with each peck value of K, it pulls back a constant specified distance above the bottom of the groove created by the previous peck to break the chip. That specified distance it pulls back is defined in Setting 22.

Haas CNC Lathe
G75 O.D. / I. D. Grooving Cycle

- **X** X-axis absolute pecking depth, diameter value.
- **Z** Z-axis absolute location to the furthest peck. (optional)
- **U** X-axis incremental pecking depth, diameter value. (optional)
- **W** Z-axis incremental distance and direction (+or-) to the furthest peck. (optional)
- **I** X-axis pecking depth increment, radius value. (optional)
- **K** Z-axis shift increment between pecking cycles. (optional)
- **D** Tool shift amount when returning to clearance plane. (optional) (see note)
- **F** Feed rate.

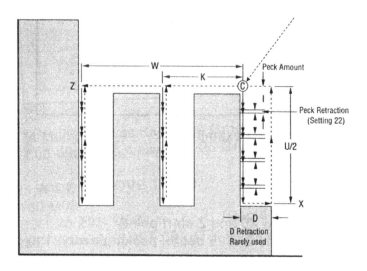

Note:
The D code command is rarely used and should only be used if the wall on the outside of groove does not exist like the diagram above shows. The D code can be used in grooving and turning to provide a tool clearance shift, in the Z-axis, before returning in the X-axis to the "C" clearance point. But, if both sides to the groove exist during the shift, then the groove tool would break. So you wouldn't want to use the D command.

Haas CNC Lathe
G75 O.D. / I.D. Grooving Cycle, Single Pass

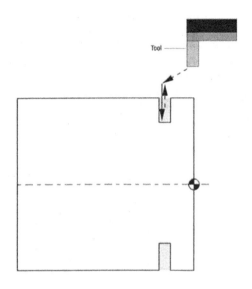

```
%
o00109
(G75 O.D. / I.D. Single pass groove cycle)
(Machine a .25" wide O.D. groove with .25" groove tool)
N1 G28
N2 T505 (.25" Wide O.D. groove tool)
N3 G97 S960 M03
N4 G54 G00 X2.1 Z0.1 M08 (Rapid to clearance point)
N5 Z-0.75 (Rapid to a start point of groove)
N6 G75 X1.75 I0.05 F0.005
(G75 Single pass O.D. grooving cycle)
N7 M09
N8 G28
N9 M30
%
```

Setting 22 (Can cycle delta Z) – As the groove tool pecks deeper into the part, with each peck value of I, it pulls back a constant specified distance above the bottom of the groove created by the previous peck to break the chip. That specified distance it pulls back is defined in Setting 22.

Haas CNC Lathe
G75 O.D. / I.D. Grooving Cycle, Multiple Pass

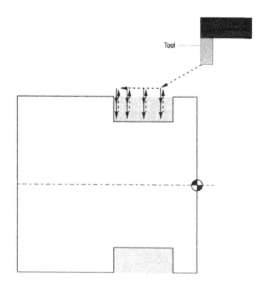

```
%
o00110
(G75 O.D. / I.D. Multiple pass groove cycle)
(Machine a 1" wide O.D. groove with .25" groove tool)
N1 G28
N2 T505 (.25" Wide O.D. groove tool)
N3 G97 S960 M03
N4 G54 G00 X2.1 Z0.1 M08 (Rapid to front of point)
N5 Z-0.75 (Rapid to a start point of groove)
N6 G75 X1.75 Z-1.5 I0.05 K0.2 F0.005
(G75 Multiple pass O.D. grooving cycle)
N7 M09
N8 G28
N9 M30
%
```

Setting 22 (Can cycle delta Z) – As the groove tool pecks deeper into the part, with each peck value of I, it pulls back a constant specified distance above the bottom of the groove created by the previous peck to break the chip. That specified distance it pulls back is defined in Setting 22.

Haas CNC Lathe
G76 Thread Cutting Cycle, Multiple Pass

- X X-axis absolute thread finish point, diameter value. (optional)
- Z Z-axis absolute distance, thread end point location. (optional)
- U X-axis incremental total distance to finish point, diameter. (optional)
- W Z-axis incremental thread length finish point. (optional)
- K Thread height, radius value.
- I Thread taper amount, radius value. (optional)
- D First pass cutting depth.
- P Thread cutting method P1-P4 (Added in software ver. 6.05)
- A Tool nose angle, no decimal with A command.
    (0° to 120°, if not used then 0° is assumed) (optional)
- F Feed rate (Threading feed rate, is the thread distance per Rev.)

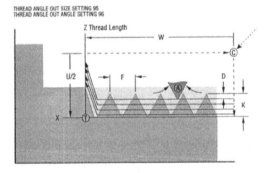

The G76 canned cycle can be used for threading both straight or tapered pipe threads. With G76 a programmer can easily command multiple cutting passes along the length of a thread.

The height of the thread is specified in K. the height of the thread is defined as the distance from the crest of the thread to the root. the calculated depth of the thread will be K less the finish allowance. Setting 86 (Thread finish allowance) is this stock allowance for a finish pass allowance, if needed.

The depth of the first cut of the thread is specified in D. this also determines the number of passes over the thread based on the value of K and the cutting method used.

The depth of the last cut on the thread can be controlled with Setting 99 (Thread minimum cut). The last cut will never be less than this value. The default value is .001 inches / .01 mm.

The feed rate is the lead of thread. The F feed rate in a G76 threading cycle is 1.0 divided by the number of threads per inch = F. (1.0 / 12 TPI = F.083333)

Haas CNC Lathe
G76 Thread Cutting Cycle, Multiple Pass

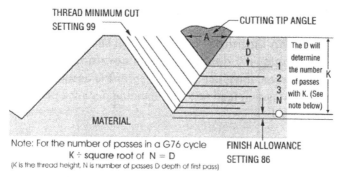

Note: For the number of passes in a G76 cycle
K ÷ square root of N = D
(K is the thread height, N is number of passes D depth of first pass)

FINISH ALLOWANCE
SETTING 86

P1: FANUC SINGLE EDGE CUTTING IN, CONSTANT ONLY

G76 Thread Cutting Cycle, Multiple Pass
Tapered Thread

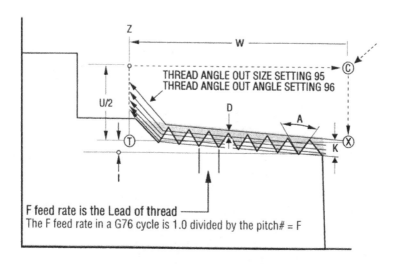

F feed rate is the Lead of thread
The F feed rate in a G76 cycle is 1.0 divided by the pitch# = F

The thread taper distance amount is specified with the I command. It is measured from the target end position in X and Z-axis down to the point in X-axis where this cycle begins and is a radius amount. A conventional O.D. taper thread will have a negative I value and a conventional I.D. taper thread will have a positive I value.

Haas CNC Lathe
G76 Multiple Pass Threading Cycle

Setup Information, Setting 33: FANUC
G76 Multiple Threading Cycle To Machine a
3/4-16 UNF-2A Thread
Tool 6, Offsets 06, Radius 0, Tip 0

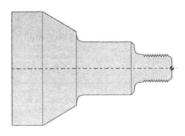

```
%
o00113
N10 G28
N20 T606
N30 G97 S720 M03
N40 G54 G00 X.85 Z1. M08
N50 Z.2 M23
N60 G76 X.674 Z-1.25 K0.0383 D.0122 F0.625
N70 M09
N80 G28
N90 M30
%
```

M23 Chamfer (angle out of thread) at end of thread is ON. An angle out of thread move can improve the appearance and functionality of a thread. This M23 commands the control to exit the thread with angle out move on a thread executed by a G76 or G92. This M code is modal and is also the default. It remains in effect until changed by M24. Refer to Settings 95 and 96 to control the move distance and angle. M23 will again be active, with an M30, Reset, or a POWER ON condition.

Setting 95 (Thread chamfer size) – The distance of angling out of the thread. The distance is designated thread pitch, so that if 1.0 is in Setting 95 and the threading feed-rate is .05, then the angle out distance will be .0500. The default in Setting 95 is 1.000.

Setting 96 (Thread chamfer angle) – Angle out of thread chamfer. The default angle of 45-degrees is in Setting 96.

M24 Chamfer (Angle out of thread) at end of thread if OFF.

An M24 commands the control to perform no angle out departure move at the end of a G76 or G92 threading cycle. This M code is modal. M24 is cancelled with an M23 (Chamfer at end of thread ON), RESET, M30 or a POWER ON condition.

Haas CNC Lathe
G76 O.D. Threading

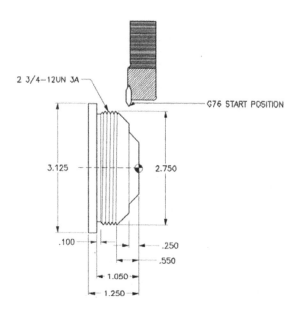

2-3/4 – 12UN 3A Thread
Major Dia. is 2.7500 / 2.7386
Minor Dia. is 2.6478
Pitch Dia. is 2.6959 / 2.6914
Thread Pitch is .083333

```
%
o00100
N1 G28 (O.D. multiple pass threading with G76)
N2 T101 (O.D. threading tool)
N3 G97 S590 M03 (Cancel CSS 590 spindle speed)
N4 G54 G00 X3.00 Z-250 M08
(Rapid X & Z above part, coolant ON)
N5 Z-.250 M24 (Start of Thread)
N6 G76 X2.6478 Z-.975 K.0511 D.0162 F.0833
(Use G76 Thread Cycle 10 Passes)
N7 M09 (Turn coolant OFF)
N8 G28 (Return to reference point)
N9 M30
%
```

Haas CNC Lathe
G81 Drill in, Rapid Out, Canned Cycle

X   Absolute X-axis rapid location. (optional)
Z   Absolute Z-depth (Feeding to Z-depth starting from R-plane) (optional)
W   Incremental Z-depth (Feeding to Z-depth starting from R-plane) (optional)
R   Rapid to R-plane (Where you rapid, to start feeding)
F   Feed-rate

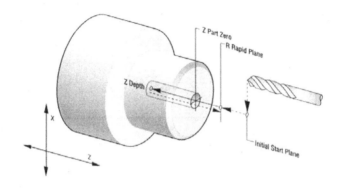

Setup: 1/2 Dia. Drill, Tool 1, Offset 01, Radius 0, Tip 0

```
%
o00119 (G81 Drilling)
N1 G28
N2 T101 (1/2 Dia. Drill) (Tool 1, Offset 1)
N3 G97 S1450 M03
N4 G54 G00 X0. Z1. M08 (Rapid to initial start point)
N5 G81 Z-0.625 R0.1 F0.005 (G81 Drilling cycle)
N6 G80 G00 Z1. M09
N7 G28
N8 M30
%
```

Haas CNC Lathe
G82 Drill In, Dwell, Rapid Out, Canned Cycle

X   Absolute X-axis rapid location. (optional)
Z   Absolute Z-depth (Feeding to Z-depth starting from R-plane) (optional)
W   Incremental Z-depth (Feeding to Z-depth starting from R-plane) (optional)
P   Dwell time at Z-depth.
R   Rapid to R-plane (Where you rapid, to start feeding)
F   Feed-rate

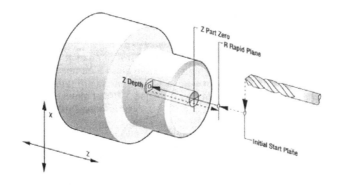

Setup: 1/2 Dia. flat bottom drill,
Tool 2, Offset 02, Radius 0, Tip 0

```
%
o00120 (G82 Drilling with a dwell)
N1 G28
N2 T101 (1/2 Dia. flat bottom drill) (Tool 2, Offset 2)
N3 G97 S1450 M03
N4 G54 G00 X0. Z1. M08 (Rapid to initial start point)
N5 G82 Z-0.625 P0.5 R0.1 F0.005
(G82 Drill with a dwell at Z-depth cycle)
N6 G80 G00 Z1. M09
N7 G28
N8 M30
%
```

Haas CNC Lathe
G83 Deep Hole Peck Drilling Canned Cycle

- X Absolute X-axis rapid location. (optional)
- Z Absolute Z-depth. (Feeding to Z-depth starting from R-plane) (optional)
- W Incremental Z-depth. (Feeding to Z-depth starting from R-plane) (optional)
- Q Pecking depth amount, always incremental. (If I, J and K are not used)
- I Size of first peck depth. (If Q is not used)
- J Amount reducing each peck after first peck depth. (If Q is not used)
- K Minimum peck depth. (If Q is not used)
- P Dwell time at Z-depth.
- R Rapid to R-plane. (Where you rapid, to start feeding)
- F Feed-rate.

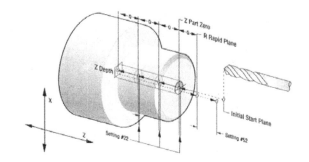

If Q is specified with the G83 command. Every pass will cut in by the Q amount, then rapid out to the R plane to clear chips and flush with coolant, then rapid into hole for the next Q peck amount until Z depth is reached.

```
%
o00121
N1 G28
N2 T303 (1/2" Dia. drill) (Tool 3, Offset 3)
N3 G97 S1820 M03
N4 G54 G00 X0. Z1. M08 (Rapid to start point)
N5 G83 Z-1.5 Q0.2 R0.1 F0.005
(G83 peck drilling cycle with Q)
N6 G80 G00 Z1.0 M09
N7 G28
N8 M30
%
```

Haas CNC Lathe
G83 Using I, J, & K To Define A Peck Amount

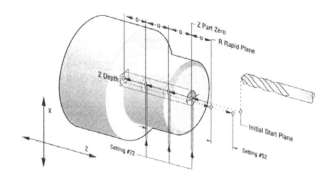

```
%
o00122
N1 G28
N2 T303 (1/2" Dia. drill) (Tool 3, Offset 3)
N3 G97 S1820 M03
N4 G54 G00 X0. Z1. M08 (Rapid to start point)
N5 G83 Z-1.5 I0.5 J0.1 K0.2 R0.1 F0.005
(G83 peck drilling cycle with I, J, K)
N6 G80 G00 Z1.0 M09
N7 G28
N8 M30
%
```

Setting 22 – As the tool pecks deeper into the hole with a G83. after each peck it rapids out to the R-plane, and then back into a constant specified distance above the bottom of the hole that was created by the previous peck. That specified distance is defined in Setting 22.

Setting 52 – Changes the way G83 works when it returns to the R-plane. Most programmers set the R-plane well above the cut to insure that the chip clear motion actually allows the chips to get out of the hole but this causes a wasted motion when first drilling through this "empty" space. Or you may need to define a clearance move above the part, in which the R-plan may be down inside a part or pocket. If Setting 52 is set to the distance required to clear chips, the R-plane can be put much closer to the part being drilled. The Z-axis will be moved above the R-plane by this amount in Setting 52.

Haas CNC Lathe
G84 Tapping Canned Cycle

- X   Absolute X-axis rapid location
- Z   Absolute Z-depth (Feeding to z-depth Starting from R-plane)
- W   Incremental Z-depth (Feeding to Z-depth starting from R-plane)
- R   Rapid to R-plane (Where you rapid, to start feeding)
- F   Feed-rate

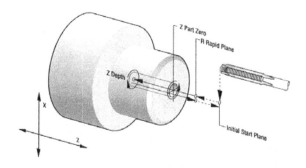

This cycle will perform different if the rigid tapping option is active. With rigid tapping, the federate for the spindle speed must be precisely the thread pitch being cut.

The feed rate is the "Lead of Thread." To calculate "F" federate in a G84 threading cycle divided 1.0 by the number of threads per inch = the "F" federate in a G84.

Example: 1.0 / 12 tpi = F.08333

You don't need to start the spindle CW before this G84 canned cycle. The control turns it on automatically.

```
%
o00123
N1 G28
N2 T404 (3/8-16 Tap)
N3 G97 S650 M05 (G84 will turn on spindle for you)
N4 G54 G00 X0. Z1. M08
N5 G84 Z-0.5 R0.2 F0.0625
N6 G80 G00 Z1.0 M09
N7 G28
N8 M30
%
```

Haas CNC Lathe
G184 Reverse Tapping, Canned Cycle

X Absolute X-axis rapid location
Z Absolute Z-depth (Feeding to z-depth Starting from R-plane)
W Incremental Z-depth (Feeding to Z-depth starting from R-plane)
R Rapid to R-plane (Where you rapid, to start feeding)
F Feed-rate

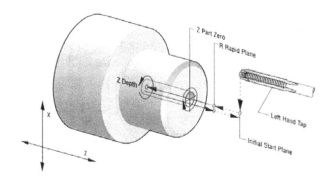

This cycle will perform different if the rigid tapping option is active. With rigid tapping, the federate for the spindle speed must be precisely the thread pitch being cut.

The feed rate is the "Lead of Thread." To calculate "F" federate in a G184 threading cycle divided 1.0 by the number of threads per inch = the "F" federate in a G184.

Example: 1.0 / 12 TPI = F.08333

You don't need to start the spindle CW before this G184 canned cycle. The control turns it on automatically.

```
%
o00124
N1 G28
N2 T404 (9/16 Left handed Tap)
N3 G97 S650 M05 (G84 will turn on spindle for you)
N4 G54 G00 X0. Z1. M08 (Rapid to initial start point)
N5 G184 Z-0.625 R0.2 F0.08333 (G184 Tapping cycle)
N6 G80 G00 Z1.0 M09
N7 G28
N8 M30
%
```

Haas CNC Lathe
G85 Bore in, Bore Out, Canned Cycle

- X Absolute X-axis rapid location (optional)
- Z Absolute Z-depth (Feeding to Z-depth starting from R-plane) (optional)
- U Incremental Z-depth location (optional)
- W Incremental Z-depth (Feeding to Z-depth starting from R-plane) (optional)
- R Rapid to R-plane (Where you rapid, to start feeding)
- F Feed-rate

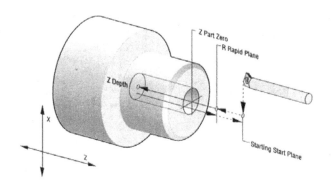

```
%
o00125
N1 G28
N2 T505 (Boring bar)
N3 G97 S1820 M03
N4 G54 G00 X0.625 Z1. M08 (Rapid to initial start point)
N5 G85 Z-0.5 R0.1 F0.005 (G85 Bore in, bore out cycle)
N6 G80 G00 Z1.0 M09
N7 G28
N8 M30
%
```

Haas CNC Lathe
G86 Bore In, Stop, Rapid Out, Canned Cycle

X Absolute X-axis rapid location (optional)
Z Absolute Z-depth (Feeding to Z-depth starting from R-plane) (optional)
U Incremental Z-depth location (optional)
W Incremental Z-depth (Feeding to Z-depth starting from R-plane) (optional)
R Rapid to R-plane (Where you rapid, to start feeding)
F Feed-rate

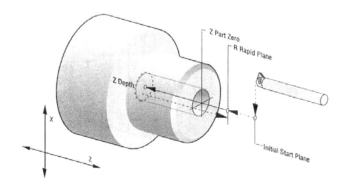

```
%
o00126
N1 G28
N2 T606 (Boring bar)
N3 G97 S1820 M03
N4 G54 G00 X0.325 Z1. M08 (Rapid to initial start point)
N5 G86 Z-0.55 R0.1 F0.005
(G85 Bore in, stop, bore out cycle)
N6 G80 G00 Z1.0 M09
N7 G28
N8 M30
%
```

Haas CNC Lathe
G87 Bore In, Manual Retract, Canned Cycle

X   Absolute X-axis rapid location (optional)
Z   Absolute Z-depth (Feeding to Z-depth starting from R-plane) (optional)
U   Incremental Z-depth location (optional)
W   Incremental Z-depth (Feeding to Z-depth starting from R-plane) (optional)
R   Rapid to R-plane (Where you rapid, to start feeding)
F   Feed-rate

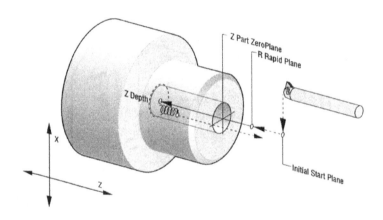

%
o00127
N1 G28
N2 T707 (Boring bar)
N3 G97 S1820 M03
N4 G54 G00 X0.25 Z1. M08 (Rapid to initial start point)
N5 G87 Z-0.625 R0.1 F0.005
(G85 Bore in, stop, manual retract cycle)
N6 G80 G00 Z1.0 M09
N7 G28
N8 M30
%

Haas CNC Lathe
G88 Bore In, Dwell, Manual Retract, Canned Cycle

X Absolute X-axis rapid location (optional)
Z Absolute Z-depth (Feeding to Z-depth starting from R-plane) (optional)
U Incremental Z-depth location (optional)
W Incremental Z-depth (Feeding to Z-depth starting from R-plane) (optional)
P Dwell time at Z-depth (optional)
R Rapid to R-plane (Where you rapid, to start feeding)
F Feed-rate

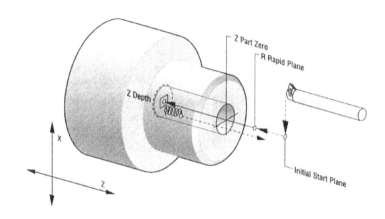

```
%
o00128
N1 G28
N2 T808 (Boring bar)
N3 G97 S1820 M03
N4 G54 G00 X0.875 Z1. M08 (Rapid to initial start point)
N5 G88 Z-0.5 P0.5 R0.1 F0.005
(G88 Bore in, dwell, bore out, manual retract cycle)
N6 G80 G00 Z1.0 M09
N7 G28
N8 M30
%
```

Haas CNC Lathe
G89 Bore In, Dwell, Bore Out, Canned Cycle

- X   Absolute X-axis rapid location (optional)
- Z   Absolute Z-depth (Feeding to Z-depth starting from R-plane) (optional)
- U   Incremental Z-depth location (optional)
- W   Incremental Z-depth (Feeding to Z-depth starting from R-plane) (optional)
- P   Dwell time at Z-depth (optional)
- R   Rapid to R-plane (Where you rapid, to start feeding)
- F   Feed-rate

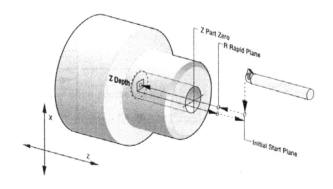

```
%
o00129
N1 G28
N2 T909 (Boring bar)
N3 G97 S1820 M03
N4 G54 G00 X0.25 Z1. M08 (Rapid to initial start point)
N5 G89 Z-0.625 P0.5 R0.1 F0.005
(G89 Bore in, dwell, bore out cycle)
N6 G80 G00 Z1.0 M09
N7 G28
N8 M30
%
```

Haas CNC Lathe
G90 O.D. / I.D. Turning Cycle

X   Absolute X-axis target location (optional)
Z   Absolute Z-axis target location (optional)
U   Incremental X-axis target distance, diameter (optional)
W   Incremental Z-axis target distance (optional)
I   Distance and direction of X-axis taper, radius value (optional)
F   Feed-rate

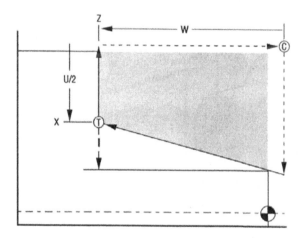

G90 is a modal canned cycle. It can be used for simple turning. Since it is modal, you can do multiple passes for turning by just specifying a new X location for successive passes.

Straight turning cuts can be made by just specifying X, Z, and F. By adding I a taper cut can be made. The amount of taper is defined with the I value added to the X value target point.

Any of the four ZX quadrants can be programmed by varying U, W, or X, and Z. The taper can be positive or negative. Selecting the sign direction is not intuitive.

Haas CNC Lathe
G90 Modal Turning Cycle With TNC

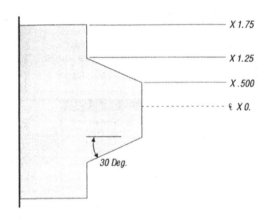

This example uses tool nose compensation with a G90 modal rough turning cycle.

```
%
o00131
N11 G28
N12 T101
N13 G50 S3000
N14 G97 S480 M03
N15 G54 G00 X1.85 Z1. M08 (Rapid to start point)
N16 G96 S390
N17 Z0.1
N18 G90 G42 X1.65 Z-0.6495 I-0.375 F0.006
(Rough 30° angle to X2.3476 Dia. using G90 and TNC)
N19 X1.55 (Additional pass)
N20 X1.45 (Additional pass)
N21 X1.35 (Additional pass)
N22 X1.25 (Additional pass)
N23 G00 G40 X3.1 Z1. M09 (TNC Departure)
N24 M05
N25 G28
N26 M30
%
```

Haas CNC Lathe
G92 Thread Cutting Cycle, Model

X Absolute X-axis target location (optional)
Z Absolute Z-axis target location (optional)
U Incremental X-axis target distance, diameter (optional)
W Incremental Z-axis target distance (optional)
I Distance and direction of X-axis taper, radius value (optional)
F Feed-rate

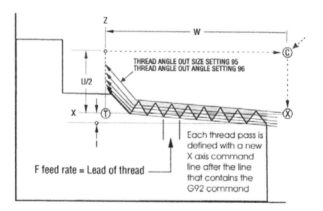

G92 is a modal canned cycle. It can be used for simple threading. Since it is modal, you can do multiple passes for threading by just specifying a new X location for successive passes.

Straight threads can be made by just specifying X, Z, and F. By adding I a pipe or taper thread can be cut. The amount of taper is defined with the I value added to the X value target point. At the end of the thread, an automatic chamfer is executed before reaching the target default for this chamfer is one thread at 45-degrees. These values can be changed with Setting 95 and Setting 96.

Any of the four ZX quadrants can be programmed by varying U, W, or X, and Z. The taper can be positive or negative. Selecting the sign direction is not intuitive. The figure shows a few examples of the values required for machining in each of the four quadrants.

Haas CNC Lathe
G92 Thread Cutting Cycle, Model

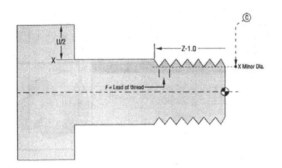

```
%
o00133
N10 (1.0-12UNF-2A)
N11 G28
N12 T404
N13 G97 S825 M03
N14 G54 G00 X1.1 Z1. M08 (Rapid to start point)
N15 Z0.25
N16 G92 X.98 Z-1.05 F0.8333 M23
(First pass of a G92 O.D. thread cycle)
N17 X.96 (Additional pass)
N18 X.94 (Additional pass)
N19 X.935 (Additional pass)
N20 X.93 (Additional pass)
N21 X.925 (Additional pass)
N22 X.9225 (Additional pass)
N23 X.92 (Additional pass)
N24 X.9175 (Additional pass)
N25 X.9155 (Additional pass)
N26 X.915 (Additional pass)
N27 X.9148 (Additional pass)
N28 G00 X1.1 Z1. M09
N29 M05
N30 G28
N31 M30
%
```

Haas CNC Lathe
Thread Cutting Chart

Threads Per Inch (TPI)	Thread Lead Feedrate in Inches Per Rev (1.0/TPI)	O. D. Thread Single Depth "K" Thread Height	I.D. Thread Single Depth "K" Thread Height	Spindle Speed Threads @ 125 IPM RPM
7	.142857	.0876	.0773	875
8	.125	.0767	.0677	1000
9	.111111	.0682	.0601	1125
10	.1	.0613	.0541	1250
11	.090909	.0558	.0492	1375
12	.83333	.0511	.0451	1500
13	.076923	.0472	.0416	1625
14	.071429	.0438	.0387	1750
16	.0625	.0383	.0338	2000
18	.055556	.0341	.0301	2250
20	.05	.0507	.0271	2500
24	.041667	.0256	.0225	3000
28	.035714	.0219	.0193	3500
32	.03125	.0192	.0169	4000
36	.027778	.0170	.0150	4500
40	.025	.0153	.0135	5000

Haas CNC SL-10 Lathe

Haas CNC Lathe
Thread Cutting Chart
"D" Value Chart for O.D. Threads

Calculated First Pass Depth ("D") For A desired Number Of Passes

Threads Per Inch

Passes	7 tpi	8 tpi	9 tpi	10 tpi
6	.0358	.0313	.0278	.0250
7	.0331	.0290	.0258	.0232
8	.0310	.0271	.0241	.0217
9	.0292	.0256	.0227	.0204
10	.0277	.0243	.0216	.0194
11	.0264	.0231	.0206	.0185
12	.0253	.0221	.0197	.0177
13	.0243	.0213	.0189	.0170
14	.0234	.0205	.0182	.0164
15	.0226	.0198	.0176	.0158
16	.0219	.0192	.0171	.0153

Threads Per Inch

Passes	11 tpi	12 tpi	13 tpi	14 tpi
6	.0228	.0209	.0193	.0179
7	.0211	.0195	.0178	.0166
8	.0197	.0181	.0167	.0155
9	.0186	.0170	.0157	.0146
10	.0176	.0162	.0149	.0139
11	.0168	.0154	.0142	.0132
12	.0161	.0148	.0136	.0127
13	.0155	.0142	.0131	.0122
14	.0149	.0137	.0126	.0117
15	.0144	.0132	.0129	.0113
16	.0140	.0128	.0118	.0110

Haas CNC Lathe:
Thread Cutting Chart
"D" Value Chart for O.D. Threads

Calculated First Pass Depth ("D") For
A desired Number Of Passes

Threads Per Inch

Passes	16 tpi	18 tpi	20 tpi	24 tpi
3	.0157	.0139	.0177	.0147
4	.0145	.0129	.0153	.0128
5	.0136	.0121	.0137	.0114
6	.0128	.0114	.0125	.0101
7	.0122	.0108	.0116	.0096
8	.0116	.0103	.0108	.0090
9	.0111	.0098	.0102	.0085
10	.0106	.0095	.0097	.0081
11	.0103	.0091	.0093	.0077
12	.0099	.0088	.0089	.0074
13	.0096	.0085	.0085	.0071

Threads Per Inch

Passes	28 tpi	32 tpi	36 tpi	40 tpi
3	.0126	.0111	.0098	.0089
4	.0110	.0096	.0085	.0077
5	.0098	.0086	.0076	.0069
6	.0089	.0078	.0070	.0063
7	.0083	.0073	.0064	.0058
8	.0077	.0068	.0060	.0054
9	.0073	.0064	.0057	.0051
10	.0069	.0061	.0054	.0049
11	.0066	.0058	.0051	.0046
12	.0063	.0055	.0049	.0044
13	.0061	.0053	.0047	.0043

Haas CNC Lathe
G94 End Face Cutting Cycle

X   Absolute X-axis target location (optional)
Z   Absolute Z-axis target location (optional)
U   Incremental X-axis target distance, diameter (optional)
W   Incremental Z-axis target distance (optional)
K   Distance and direction of Z-axis coning (optional)
F   Feed-rate

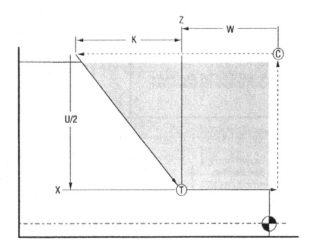

G94 is a modal canned cycle. It can be used for simple end facing. Since it is modal, you can do multiple passes for facing by just specifying a new Z location for successive passes.

Straight end facing cuts can be made by just specifying X, Z, and F. By adding K a conical face can be cut. The coning amount is defined with the K value that is added to the Z value target point.

Any of the four ZX quadrants can be programmed by varying U, W, or X, and Z. The coning can be positive or negative. Selecting the sign direction is not intuitive.

Haas CNC Lathe
G94 Modal End Facing With TNC

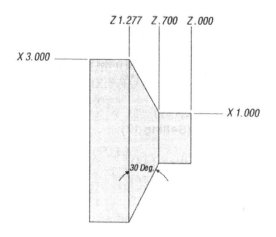

This example uses tool nose compensation with a G94 modal rough facing cycle.

```
%
o00135
N11 G28
N12 T101 (O.D. Facing tool)
N13 G50 S3000
N14 G97 S480 M03
N15 G54 G00 X3.1 Z1. M08 (Rapid above part)
N16 G96 S390
N17 Z.1 (Rapid to start point)
N18 G94 G41 X1.0 Z-0.3 K-0.5774 F0.01
(Rough 30° angle to X1. and Z-0.7 using G94 and TNC)
N19 Z-0.4 (Additional pass)
N20 Z-0.5 (Additional pass)
N21 Z-0.6 (Additional pass)
N22 Z-0.69 (Additional pass)
N23 Z-0.7 (Additional pass)
N24 G40 G00 X3.1 Z1. M09 (Cancel TNC)
N25 M05
N26 G28
N27 M30
%
```

Haas CNC Lathe
Live Tooling

This Option is not field installable. The live tooling option allows the user to drive VDI axial or radial tools to perform such operations as milling, drilling or slotting. The main spindle of the lathe is indexable in one-degree increments for precise part positioning and repeatability. Milling shapes is possible using spindle motion G codes.

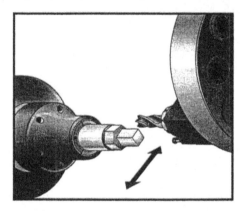

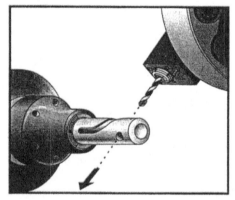

Programming Notes:

The live tool drive will automatically turn itself off when a tool change is commanded. The main spindle can be clamped (M14, M15) for using the live tooling. It will automatically unclamp when a new main spindle speed is commanded or reset is pressed. Haas live tooling is designed for medium duty milling, a 3/4" diameter end mill in mild steel is the maximum. And the maximum live tooling drive speed is 3000 rpm.

- Canned cycles are not supported.
- Large tool diameters may require reduction tool holders.

Haas CNC Lathe
Live Tooling, M Codes

M19 Angle CMD (optional): An M19 will orient the spindle to the zero position. A "P" value can be added that will cause the spindle to orient to a specific position in degrees of accuracy "P" rounds to the nearest whole degree, and "R" rounds to the nearest hundredth of a degree (x. xx) when the angle is viewed in the current commands tool load screen.

M133 Live Tool Drive Forward:
M134 Live Tool Drive Reverse:
M135 Live Tool Drive Stop:

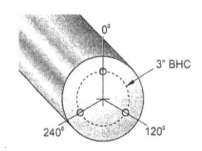

Program Example:
G00 X3.0 Z0.1
G98
M19 P0
G04 P2. (dwell time for motor stabilization)
M14
M133 P2000
G01 Z-0.5 F40.0
G00 Z0.1
M19 P120
G04 P2. (dwell time for motor stabilization)
M14
G01 Z-0.5
G00 Z0.1
M19 P240
G04 P2. (dwell time for motor stabilization)
M14
G01 Z-.05
G00 Z0.1
M15
M135

Haas CNC Lathe
Live Tooling, Synchronous Milling

G32 synchronous motion is a control mode where the X, Z-axis are commanded to move distances at constant feed-rates and spindle is commanded to rotate at constant speeds.

G32 is commonly used to create threads, the spindle rotates at a constant rpm and constant Z-axis motion begins at the same reference Z-axis mark for each stroke. Many strokes can be repeated because the reference mark sets the location of the start thread.

Geometric shapes can be machined using G32, however, G32 motions can be cumbersome to create and difficult to adjust in the final program. To relieve the user of this burden, the Haas CNC control has a canned cycle G code which simplify the creation of simple geometric shapes. G77 flatting cycle automates the motions of 1 or more sided uniform shapes.

In addition to synchronous motions G5 is a motion mode that accepts point-to-point commands and controls the spindle like a rotary device, similar to a rotary table motion. It is commanded in angle and distance point to point motion.

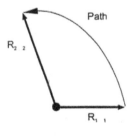

G32 paths between commanded points are curves.

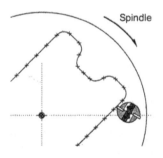

Using, many small motion commands can result in geometric shapes.

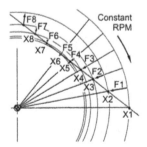

G32 motion includes both X feed-rate and Position commands at a constant RPM.

Haas CNC Lathe
Live Tooling, Fine Spindle Control

Many uses of live tooling involve holding the spindle still while performing a cut with the live tool. For certain types of operations, it is necessary to move this spindle in a controlled manner while cutting with the live tool

Fine spindle control (FSC) is most commonly used to create features on or near the face of a part, such as grooves, slots, and flat surfaces. Typically an end-mill pointing along the Z-axis is used to perform the cutting, after pilot holes are drilled. Live tooling is almost always required in order to use FSC. Single point turning is not recommended as the surface feed per minute required is to high for the FSC function.

Limitations: The primary function of the spindle is to turn rapidly. The introduction of G codes for FSC dose not change the mechanical design of the spindle motor. Therefore, you should be aware of certain factors that apply when the spindle is turning at very low torque. This limits the depth of cut that can practically be performed with the live tool while the spindle is not locked. In many cases you will want to track the motion of the spindle with motion in the X-axis.

The limit also applies to positioning the spindle in general. This has an effect when trying to perform cuts that are close to centerline. The number of control points depends on the radius and direction of cutter path. Cutter paths with a large radius and a shallow angle towards the center will result in few control points.

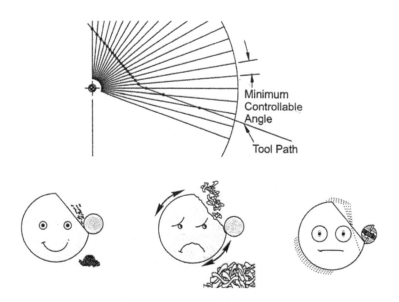

Light Cut, Good Finnish   Heavy Cut, Bad Finnish   Large Cutter, Chatter, Bad Finnish

Haas CNC Lathe
Live Tooling, Installation

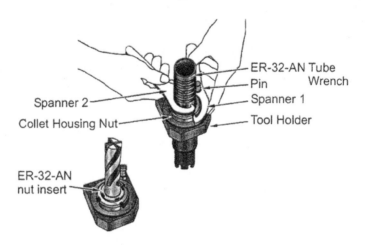

Insert the tool bit into the ER-AN nut insert. Thread the nut insert into the collet-housing nut.

Place the ER-32-AN tube wrench over the tool bit and engage the teeth of the ER-AN nut insert. Snug the ER-AN nut insert by hand using the tube wrench.

Place Spanner 1 over the pin and lock it against the collet-housing nut. It may be necessary to turn the collet-housing nut to engage the spanner.

Engage the teeth of the tube wrench with Spanner 2 and tighten the nut.

De Anza Manual Machine Lab

Haas CNC Lathe
Live Tooling, VDI Adapter Installation

VDI adapters make it possible to use VDI-40 tools in Haas lathe turrets. Installation Procedure:

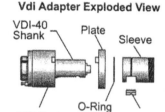

Vdi Adapter Exploded View

Install plate over VDI-40 shank. Orient plate boss to VDI tool counterbore.

Slide adapter sleeve onto tool shank with cut-out facing towards base of tool shank. Align cut-out with tooth profile of shank.

Insert key into sleeve cut-out. Ensure tooth profile of key fits into tool shank properly.

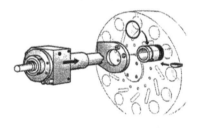

Vdi Adapter Installation

Place O-ring in the groove as shown. The O-ring will keep the key from falling out.

Install VDI tool with adapter into turret. Ensure the turret locating pin and plate hole are properly aligned.

Tighten the draw nut to lock assembly in place.

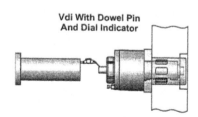

Vdi With Dowel Pin And Dial Indicator

Live Tooling, C-Axis: This operation provides high-precision bi-directional spindle motion that is fully interpolated with X and/or Z motion. Spindle speeds from .01 to 60 rpm can be commanded.

Haas CNC Lathe
Live Tooling

C-Axis:

This operation provides high-precision bi-directional spindle motion that is fully interpolated with X and/or Z motion. Spindle speeds from .01 to 60 rpm can be commanded.

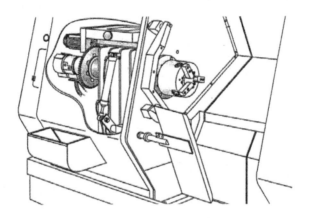

C-Axis Operation:

M154 C-axis engage

M155 C-axis Disengage

Setting 101 Diameter used to calculate the feed-rate.

The lathe will automatically disengage the spindle brake when the C-axis is commanded to move and to reengage it afterwards, if it has previously been engaged.

C-Axis incremental moves are possible using the "H" address code as shown in the following example.

G0 C90. (C-Axis moves to 90. degrees)

H-10 (C-Axis moves to 80. degrees)

Haas CNC Lathe
Live Tooling, Cartesian to Polar

Cartesian to Polar Transformation: Cartesian to Polar coordinate programming that converts X, Y position commands into rotary C-axis and linear X-axis moves. Cartesian to Polar coordinate programming reduces the amount of code required to; command complex moves. Normally a straight line would require many points to define the path, while in Cartesian, only the end points are necessary. This feature allows face machining programming in the Cartesian coordinate system.

Programming Notes: Programmed moves should always position the tool centerline.

Do not use G41 or G42 cutter compensation.

Tool paths should never cross the spindle centerline. Cuts that must cross spindle center can be accomplished with two parallel passes on either side of spindle center.

Cartesian to Polar conversion is a model command (see the G-code section)

Z-axis moves are not allowed while this mode is enabled.

Cartesian Interpolation: Cartesian coordinate commands are interpreted into movements of the linear axis, (turret movements) and spindle movements, (rotation of the workpiece).

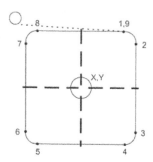

(Example Program)
%
o00069
N6 (square)
G59
(tool 11. 3/4" diameter end-mill, cutting on center)
T1111
M154 G00 C0.
G97 M133 P1500
G00 Z1.
G00 G98 X2.35 Z0.1 (position)
G01 Z-0.05 F25.
G112
G17

(Example Continued)
G0 X-.75 Y.5
G01 X0.45 F10. (point 1)
G02 X0.5 Y0.45 R0.05 (point 2)
G01 Y-0.45 (point 3)
G02 X0.45 Y-0.5 R0.05 (point 4)
G01 X-0.45 (point 5)
G02 X0.5 Y-0.45 R0.05 (point 6)
G01 Y0.45 (point 7)
G02 X-0.45 Y0.5 R0.05 (point 9)
G01 X0.45 (point 9) Y.6
G113
G00 Z3.
M30
%

Haas CNC Lathe
Live Tooling, G05 Sample Programs

Example #1

M154
G00 G98 (feed/min) X2.0 Z0
G90
G01 Z-0.1 F6.0
X1.0
C180. F10.0
X2.0
G00 Z0.5
M155

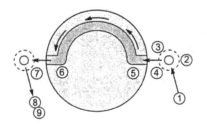

Example #2
G05 Face Slot Example: Assume pilot hole is already drilled.

N1 T303 (small end-mill)
N2 M19 (orient spindle)
N3 G00 Z0.5
N4 G00 X1.
N5 M133 P1500
N6 G98 G01 F10. Z-.25
(plunge into pre-drilled hole)
N7 G05 R90. F40. (make slot)
N8 G01 F10. Z0.5 (retract)
N9 M135
N10 G99 G28 U0 W0

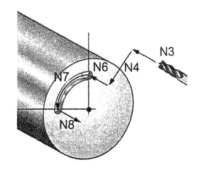

Example #3
G05 Simple Cam

N1 T303 (small end-mill)
N2 M19 (orient spindle)
N3 G00 Z-0.25
N4 G00 X2.5 (approach 2" Dia. stock)
N5 M133 P1500
N6 G98 G01 X1.5 F40
(cut to top of cam)
N7 G05 R215. X.5 F40. (cut cam)
N8 G01 X2.5 F40. (cut out of cam)
N9 M135
N10 G99 G28 U0 W0

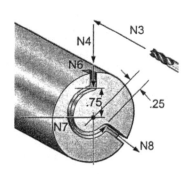

Haas CNC Lathe
Live Tooling, G05 Sample Programs

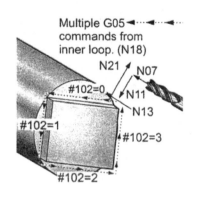
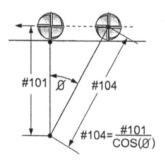

Example #4, G05 Flatting

```
%
o01484 (cut a square with G05)
N1 G28 X0.
N2 G28 Z0.
N3 G54 G00 G40 G97
N4 G103 XP3
N5 T707
N6 M19
N7 G00 Z0.5
    ()
M8 #101 = [0.707 + 0.75]
(closest approach)
(101 = center to side plus)
(half of tool diameter)
N9 #101 = #101 x 2
(multiply by 2 for diameter)
N10 #104 = [#101 / cos[45.]]
(104 = distance at corner)
N11 G98 G01 X#104 F100.
N12 M133 P1500
N13 Z-0.1
(feed into pre-drilled hole)
N14 #102 = 0
    WHILE [#103 LT 45.] DO1
(four sided shape)
N15 #103 = -45. (angle from center of flat)
    WHILE [#103 LT 45.] DO2
N16 #103 = [#103 +5.]
N17 #104 = [#101 / cos[#103]]
N18 G05 X#104 R5. F20.
    END2
N19 #102 = [#102 + 1]
    END1
    ()
N20 M135
N21 G28 U0
N22 G28 W0
N23 M30
%
```

Haas CNC Lathe
Live Tooling, G77 Sample Programs

G77 Flatting Cycle: this cycle is only available on lathes with the live tooling option.

I   Angle of first flat, in degrees (optional).
J   Distance from center to flat.
L   Number of flat surfaces to cut (optional).
R   Tool radius
S   Spindle speed (optional)
K   Part diameter (optional)

G77 with L specified

G77 with K specified

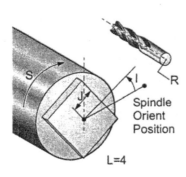

L=4

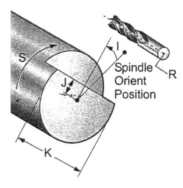

G77 operates in one of two modes, depending on whether a K code or an L code is specified. If a K code is specified, one flat surface will be cut. If L code is specified, L flat surfaces will be cut, equally spaced all the way around the part. L must be grater than or equal to 3. If two sides are desired, perform two K cuts at I angle spacing.

The J value specifies the distance from the center of the part to the center of a flat surface. Specifying a larger distance will result in a shallower cut. This may be used to perform separate roughing and finishing passes. When using an L code, care should be taken to verify that the corner to corner size of the resulting part is not smaller than the diameter of the original part, or the tool may crash into the part during its approach.

The L value allows a part with multiple flat surfaces to be specified. For example, L4 specifies a square, and L6 specifies a hex.

The S value specifies the rpm speed that the spindle will maintain during the flatting cycle. The default value is 6.

Higher values will not affect the flatness, but will affect the position of the flats. To calculate the maximum error in degrees, use RPM x .006

The I value specifies the offset of the center of the first flat surface from the zero position, in degrees. If the I value is not used, the first flat surface will start at the zero position. This is equivalent to specifying an I equal to half the number of degrees covered by the flat surface. For example, a square cut without an I value would be the same as a square cut with I set to 45.

Haas CNC Lathe
Live Tooling, G77 Sample Programs

Continued from previous page

G77 Flatting:
Cut a half-inch deep flat into the top inch of a part that is 4" in diameter, using a tool 1" in diameter.

...
N100 S10 M03 (start spindle)
N101 M133 P1000 (turn on live tool)
N102 G00 X6.1
N103 Z-1.
N104 G77 J1.5 K4. R0.5
N105 Z1.
N106 M135 (stop live tool)
N107 M05 (stop spindle)
...

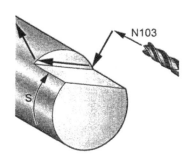

G77 Hex: Cut a hexagon into the top half inch of a part that is 3" in diameter, using a tool 1/2" in diameter.

...
N200 S10 M03 (start spindle)
N201 M133 P1000 (turn on live tool)
N202 G00 X4.5
N203 Z-0.05
N204 G77 J1.299 L6 R.25
N205 Z1.
N206 M135 (stop live tool)
N107 M05 (stop spindle)
...

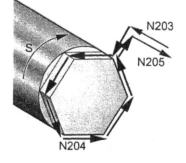

Haas CNC Lathe: Live Tooling
Live Tooling, G77 Sample Programs:

```
%
o00015 (sample 2 sided flat program)
N100 T606
N110 G97 S3 M03
N120 M133 P2000
N130 G00 X4. Z0.05
N140 Z-1.849
N150 G77 J0.625 I0 R0.25 K2.
(J = 1.25 flat dia, I0 = flat center)
(R.25 = .5 Dia. end-mill, K = part stock Dia.)
N160 G77 J0.625 I180. R0.25 K2.
(J = 1.25 flat dia, I180. = flat center)
(R.25 = .5 Dia. end-mill, K = part stock Dia.)
N170 G00 Z1.
N180 M135
N190 M05
N200 G00 X10. Z12.
N210 M30
%
```

G77 Two Flats:
Cut a 3/8" flat into the top and bottom of a part that is 2" in diameter, using a 1/2" Diameter tool.

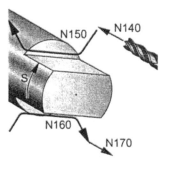

De Anza, Manual Machine Lab

Haas CNC Lathe
Live Tooling, G112 Sample Programs

G112 XY to XC Interpretation:

The G112 Cartesian to Polar coordinate transformation feature allows the user to program subsequent blocks in Cartesian XY coordinates, which the control automatically converts to polar XC coordinates. While it is active, G18 ZX plane is used for G01 linear XY strokes and G17 YZ plane is used for G02 and G03 XY interpolated circular motion.

Note:

That mill-style Cutter Compensation becomes active when G112 is used. Cutter Compensation G41 and G42 must be canceled with G40 before the next G113.

G113 cancels the Cartesian to Polar coordinate conversion.

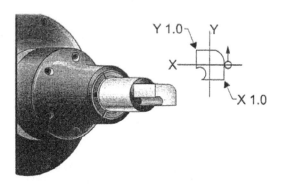

```
%
T0101
G54
G17
G12
M154
G00 G98 Z.1
G00 X.875 Y0.
M08
G97 P2500 M133
G01 Z0. F15.
Y.5 F5.
G03 X.25 Y1.125 R.625
G01 X-.75
G03 X-.875 Y1. R.125
G01 Y-.25
G03 X-.75 Y-.375 R.125
G02 X-.375 Y-.75 R.375
G01 Y-1.
G03 X-.25 Y-1.125 R.125
G01 X.75
G03 X.875 Y-1. R.125
G01 Y0.
G00 Z.1
G113
G18
M9
M155
M135
G28 U0.
G28 W0. H0.
M30
%
```

Haas CNC Lathe
Live Tooling, G95 & 186

G95 Live Tool Rigid Tapping (Face)
F   Feed-rate.
R   Position of the R plane.
W   Z-axis incremental distance.
X   Optional Part Diameter
    X-axis motion command.
Z   Position of bottom of hole.

G186 Reverse Live Tool Rigid Tapping (Face)
F   Feed-rate.
R   Position of the R plane.
W   Z-axis incremental distance.
X   Optional Part Diameter
    X-axis motion command.
Z   Position of bottom of hole.

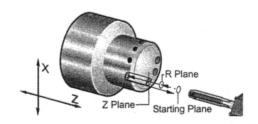

Live Tooling, G195 & 196:

G195 Live Tool Radial Tapping
F   Feed-rate.
R   Position of the R plane.
U   X-axis incremental distance.
X   X-axis motion command.
Z   Position of bottom of hole.

G196 Reverse Live Tool Victor Tapping
F   Feed-rate.
R   Position of the R plane.
U   X-axis incremental distance.
X   X-axis motion command.
Z   Position of bottom of hole.

```
%
o00800
(G195 Live Tool radial tapping)
N1 T101 (radial 1/4-20 tap)
G99 (necessary for this cycle)
G00 Z0.5
X2.5
Z-0.7
S500 (rpm should look like this,
CW direction)
M19 PXX (orient spindle at desired location)
M14 (lock spindle)
G195 X1.7 F0.05 (thread down to X1.7)
```

```
G28 U0
G28 W0
M135 (stop live tooling spindle)
M15 (unlock spindle)
M30
%
```

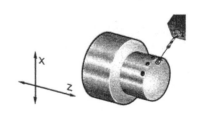

Haas CNC Lathe
Safety Procedures

Professional Machine Operators
Always Consider
Safety First

Haas CNC Lathe
Safety Procedures
Read Before Operating This Machine

Only authorized personnel should work on this machine. Untrained personnel present a hazard to themselves and the machine and improper operation will void the warranty.

Check for damaged parts and tools before operating the machine. Any part or tool that is damaged should be properly repaired or replaced by authorized personnel. Do not operate the machine if any component does not appear to be functioning correctly. Contact your shop supervisor.

Use appropriate eye and ear protection while operating the machine. ANSI-approved impact safety goggles and OSHA-approved ear protection is recommended to reduce the risks of sight damage and hearing loss.

Do not operate the machine unless the doors are closed and the door interlocks are functioning properly. Rotating cutting tools can cause severe injury. When a program is running, the mill table and spindle head can move rapidly at any time in any direction.

The Emergency Stop button (also known as an Emergency Power Off button) is the large circular red switch located on the Control Panel. Pressing the Emergency Stop button will instantly stop all motion of the machine, the servo motors, the tool changer, and the coolant pump. Use the Emergency Stop button only in emergencies to avoid crashing the machine.

The electrical panel should be closed and the key and latches on the control cabinet should be secured at all times except during installation and service. At those times, only qualified electricians should have access to the panel.

When the main circuit breaker is on, there is high voltage throughout the electrical panel (including the circuit boards and logic circuits) and some components operate at high temperatures. Therefore, extreme caution is required. Once the machine is installed the control cabinet must be locked and the key available only to qualified service personnel.

Consult your local safety codes and regulations before operating the machine. Contact you dealer anytime safety issues need to be addressed.

DO NOT modify or alter this equipment in any way. If modifications are necessary, all such requests must be handled by Haas Automation, Inc.. Any modification or alteration of any Haas Milling or Turning Center could lead to personal injury and/or mechanical damage and will void your warranty.

It is the shop owner's responsibility to make sure that everyone who is involved in installing and operating the machine is thoroughly acquainted with the installation operation and safety instructions provided with the machine BEFORE they perform any actual work. The ultimate responsibility for safety rests with the shop owner and the individuals who work with the machine.

Haas CNC Lathe
Safety Procedures
Observe All Of The Warnings And Cautions Below

Follow all of the warnings of the chuck manufacturer regarding the chuck and work holding procedures.

Hydraulic pressure must be set correctly to securely hold the work piece without distortion.

The electrical power must meet the specifications in this manual. Attempting to run the machine from any other source can cause severe damage and will void the warranty.

DO NOT press POWER UP/RESTART on the control panel until after the installation is complete.

DO NOT attempt to operate the machine before all of the installation instructions have been completed.

NEVER service the machine with the power connected.

Improperly clamped parts at high velocity may puncture the safety door. Reduced rpm is required to protect the operator when performing dangerous operations (e.g. turning oversized or marginally clamped parts). Turning oversized or marginally clamped parts is not safe.

Windows must be replaced if damaged or severely scratched -Replace damaged windows immediately.

Do not process toxic or flammable material. Deadly fumes can be present. Consult material manufacturer for safe handling of material by- products before processing.

Do not operate with the door open.

Do not operate without proper training.

Always wear safety goggles.

The machine is automatically controlled and may start at any time.

Improperly or inadequately clamped parts may be ejected with deadly force.

Do not exceed rated chuck rpm.

Higher rpm reduces chuck clamping force.

Unsupported bar stock must not extend past draw-tube end.

Chucks must be greased weekly and regularly serviced.

Chuck jaws must not protrude beyond the diameter of the chuck.

Do not machine parts larger than the chuck.

Haas CNC Lathe
Safety Procedures
Observe All Of The Warnings And Cautions Below

Unattended Operation:
Fully enclosed Haas CNC machines are designed to operate unattended; however your machining process may not be safe to operate unmonitored.

As it is the shop owner's responsibility to set up the machines safely and use best practice machining techniques, it is also their responsibility to manage the progress of these methods. The machining process must be monitored o prevent damage if a hazardous condition occurs.

For example if there is the risk of fire due to the material, then an appropriate fire suppression system must be installed to reduce the risk of harm to personnel, equipment and the building. A suitable specialist must be contacted to install monitoring tools before machines are allowed to run unattended.

It is especially important to select monitoring equipment that can immediately perform an appropriate action without human intervention to prevent an accident, should a problem be detected.

De Anza, Manual Machine Lab

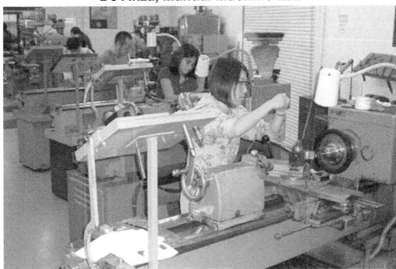

Haas CNC Lathe Specifications

Lathe	Mini	SL-10	SL-20	TL-15
Capacity				
Chuck Size	5C Collet	6.5"	8.3"	8.3"
Std. Bar Capacity	1.06"	1.75"	2.0"	2.0"
Between Centers	N / A	16.5"	24.0"	12.7"
Max. Cutting Dia.		11.0"	10.3"	8.2"
Max. Cutting Length.		14.0"	20.0"	17.5"
Spindle				
Peak Horsepower	7.5 Hp	15 Hp	20 Hp	20 Hp
Max. RPM	6,000 RPM	6,000 RPM	4,000 RPM	4,000 RPM
Spindle Nose	A2 – 5	A2 – 5	A2 – 6	A2 – 6
Spindle Bore Dia.	1.38"	2.31"	3.00"	3.00"
Draw Tube Bore Dia.	1.18"	1.81"	2.06"	2.06"
Swing Diameter				
Over Front Apron		16.25"	23.0"	20.0"
Over Cross Slide	5.8"	8.0"	9.5"	9.5"
Travels				
X – Axis	12"	6.25"	8.45"	7.5"
Z – Axis	12"	14.0"	20.0"	17.5"
Feed Rates				
Axis Motor Max Thrust	2000 Lbs	3400 Lbs	3400 Lbs	3400 Lbs
Rapid – AC Brushless	600 IPM	710 IPM	710 IPM	710 IPM
Tools				
Number Of Tools	Gang Style	Turret	Turret	Turret
	Up to 10	12	10	12
Miscellaneous				
Coolant (U.S. Gal)	24	15	40	40
Tailstock				
Taper	N / A	MT3	MT4	N / A
Travel	N / A	4.0"	20"	N / A
Thrust	N / A	225 – 900 Lbs	300 – 1500 Lbs	N / A
Live Tooling Lath	N / A	N / A	N / A	Live Tooling

Haas CNC Lathe Specifications

Lathe	SL-30	TL-25	SL-40
Capacity			
Chuck Size	10.0"	10.0"	15.0"
Std. Bar Capacity	3.0"	3.0"	4.0"
Between Centers	39.0"	39.0"	51.0"
Max. Cutting Dia.	17.0"	16.0"	25.5"
Max. Cutting Length.	34.0"	34.0"	44.0"
Spindle			
Peak Horsepower	30 Hp	30 HP	40 Hp
Max. RPM	3,400 RPM	3,400 RPM	2,400 RPM
Spindle Nose	A2-6	A2-6	A2-8
Spindle Bore Dia.	3.50"	3.50"	4.62"
Draw Tube Bore Dia.	3.03"	3.03"	4.06"
Swing Diameter			
Over Front Apron	30.0"	30.0"	40.0"
Over Cross Slide	14.5"	14.5"	25.0"
Travels			
X – Axis	11.3"	11.3"	17.0"
Z – Axis	34.0"	23.0"	44.0"
Feed Rates			
Axis Motor Max Thrust			
(X)	3,400 Lbs	3,400 Lbs	3,400 Lbs
(Z)	5,400 Lbs	5,400 Lbs	5,400 Lbs
Rapid – AC Brushless	710 IPM	710 IPM	710 IPM
Tools	Turret	Turret	Turret
Number Of Tools	12	12	10
Miscellaneous			
Coolant (U.S. Gal)	50	50	77
Tailstock			
Taper	MT4	N / A	MT5
Travel	33.5"	N / A	44"
Thrust	300 – 1,500 Lbs	N / A	300 – 1,500 Lbs
Live Tooling Lath	N / A	Live Tooling	N / A

Haas CNC Lathe
Quick Start-Up Guide

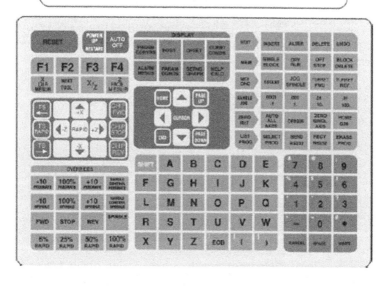

Haas CNC Lathe
Quick Start-Up Guide

The purpose of this quick-reference guide is to assist the operator in setting up and running a Haas lathe in as little time as possible. Refer to the Operators Manual for more programming and operating information.

IMPORTANT!! Be sure All installation procedures have been completed before operating the machine.

Note: DO NOT attempt the following procedures until they are first demonstrated by the Instructor.

POWERING ON THE MACHINE:
Press the green POWER ON button in the upper left hand corner of the control.

With the front door closed, press the POWER UP / RESTART button. This will return all available axes.

CREATING A PROGRAM:
Press the LIST PROG key. This will show a list of any programs presently in the control.

Press the O (letter, not number) key, then nnnnn, where nnnnn can be any five numbers. Then press the WRITE / ENTER key. This will create a program named Onnnnn. Program names must follow this format.

The new program may now be created on the line following the program number. All further entries are made by typing a letter followed by a numeric value and pressing the
WRITE / ENTER, INSERT or ALTER key. To make changes to an existing program, use the following keys:

The INSERT key will insert any text to the right of the current cursor location.

The ALTER key will replace highlighted information with the desired text.

The DELETE key will delete any highlighted text.

INDEXING A TOOL IN THE TURRET:
Press the MDI button. Type "T1", then press the TURRET FWD or TURRET REV key.

Or, When in JOG mode on the Offset display, "T1", then press NEXT TOOL.

Haas CNC Lathe
Quick Start-Up Guide
Toughing Off Part To Get Tool Offsets

Be sure to HAND JOG The machine turret away to provide clearance for a tool change.

Press the OFFSET key until the TOOL GEOMETRY page appears, and cursor to the desired tool offset number.

Type in T#, and press NEXT TOOL to index the turret to the desired tool.

Hand-jog the tool to a point on the outside of the part and touch off the tool to the part.

Move the tool away in the Z-axis only.

Press X DIA MESUR to enter the present location of the tool.

The prompt "ENTER DIM" will then be flashing in the lower left corner.

Make sure the spindle is OFF before measuring the part diameter.

Enter the part diameter, and press WRITE/ENTER for the X geometry offset.

Hand-jog the tool and touch off at a point on the face of the part to define as Z zero.

Press Z FACE MUSUR to record the Z geometry offset.

Enter any additional offset amount, if needed, and press WRITR/ENTER.

Repeat these steps for the remaining tools.

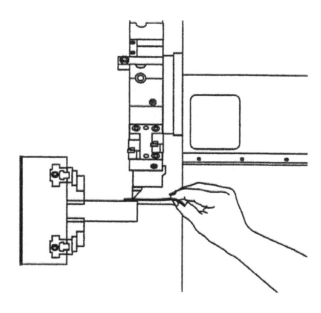

Haas CNC Lathe
Quick Start-Up Guide
Touching Off The Probe To Get Tool Offsets

Setting Tool Geometry Offsets using a Probe:

Hand-jog the machine axes to a safe tool change location and lower the probe.

Press the OFFSET key until the TOOL GEOMETRY page appears.

Cursor to the desired tool offset number.

Type in T#, and press NEXT TOOL to index the turret to the desired tool.

HANDLE near probe (hand wheeling into probe stops turret, but won't enter valve).

Using X-axis jog keys, carefully jog at 1.0 IPM until the tool tip touches probe and beeps.

Note: The control will beep, enter an offset value, and not allow the operator to continue in that direction.

To enter the X-axis offset values for center-drills, taps, and reamers, press F2.

To enter Z-axis offset, jog the Z-axis 1.0 IPM into the probe until tip touches probe and beeps.

The distance from the face of the probe to the face of the part needs to be added, using a work zero offset, for each setup that touches off using a probe.

Touch off all the tools to the probe before going to the WORK ZERO OFFSET display.

Hand-jog the turret to a safe location and raise the arm into place.

Select a tool, and hand-jog to the surface where the part program zero-point is defined.

Press the OFFSET key until the WORK ZERO OFFSET page appears. Cursor to the desired work offset number (usually G54).

Press the Z FACE MESUR key, and a work zero offset value will be entered. The machine will shift all the tools by the value of the work offset.

Haas CNC Lathe
Quick Start-Up Guide

Setting Work Zero Offsets Using A Probe:

The distance from the face of the probe to the face of the part needs to be added, using a work zero offset, for each setup that touches off using a probe.

Touch off all the tools to the probe before going to the WORK ZERO OFFSET display.

Hand-jog the turret to a safe location and raise the arm into place.

Select a tool, and hand-jog to the surface where the part program zero-point is defined.

Press the OFFSET key until the WORK ZERO OFFSET page appears. Cursor to the desired work offset number
(usually G54).

Press the Z FACE MESUR key, and a work zero offset value will be entered. The machine will shift all the tools by the value of the work offset.

Running A Program:

Press the LIST PROG key. The program list will include the program number and the first text comment in parentheses in the first two lines of a program. Cursor to the desired program and then press the SELECT PROG or WRITE / ENTER key.

Press the MEM key.

Press CYCLE START to begin running the program.

To start a program at a tool other than the beginning tool, cursor down to the tool you wish to start with, and press CYCLE START.

To start a program other than at a tool change, turn ON Setting 36, cursor down to the desired starting place in the program, and then press the CYCLE START button.

The GRAPHICS function gives you a visual dry run of your part program without moving the axis and risking tool damage from programming errors. To select GRAPHICS mode, press MEM or MDI and the SETNG / GRAPH key twice.

Haas CNC Lathe
Chuck Installation

Note: if necessary, install an adapter plate before installing the chuck.

Clean the face of the spindle and the back face of the chuck.

Position the drive dog at the top of the spindle.

Remove the jaws from the chuck.

Remove the center cup or cover-plate from the front of the chuck.

If available, install a mounting guide into the drawtube and slide the chuck over it.

Orient the chuck so that one of the guide holes are aligned with the drive dog.

Use the chuck wrench to thread the chuck onto the drawtube.

Screw the chuck all the way onto the drawtube and back it off 1/4 turn.

Align the drive dog with on of the holes in the chuck.

Tighten the six (6) socket head cap screws and install the center cup or plate with three (3) socket head cap screws.

Install the jaws. If necessary replace the rear cover plate. This is located on the left side of the machine.

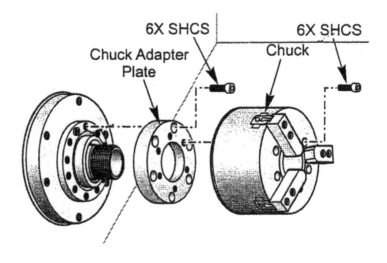

Haas Lathe
Collet Installation

Thread the collet adapter into the drawtube.

Place the spindle nose on the spindle and align one of the holes on the back of the spindle nose with the drive dog.

Fasten the spindle nose to the spindle with six (6) socket head cap screws.

Thread the collet onto the spindle nose and align the slot on collet with the set screw on the spindle nose. Tighten the setscrew on the side of the spindle nose.

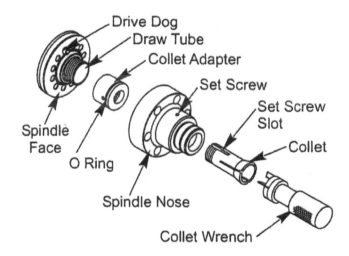

Drawtube Cover Plate

It is necessary to remove the cover plate at the far end of the drawbar when using a bar-feeder. Replace the cover plate anytime bar stock is not being fed automatically.

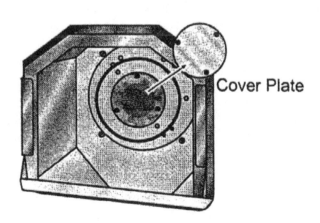

Haas CNC Lathe
Re – Positioning Chuck Jaws

Re-position chuck jaws when the jaw stroke travel cannot generate sufficient clamp force to hold the material, when changing to a smaller diameter stock.

The part will not be sufficiently clamped if there is not extra stroke before bottoming out the jaws.

Use a hex key to loosed two socket head cap screws attaching the jaw to the chuck.

Slide jaw to new position and re-tighten the two socket head cap screws.

Repeat procedure for remaining two jaws. Jaws must remain concentric.

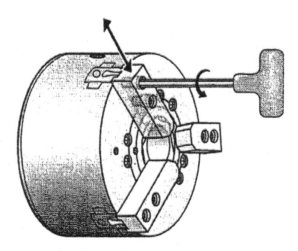

Auto Air Jet (optional)

The M12 and M13 codes are used to activate the optional Auto Air Jet. M12 turns the air blast on and M13 turns the air blast off. Additionally, M12 Pnnn (nnn in milliseconds) will turn it on for the specified time, and then turn off automatically.

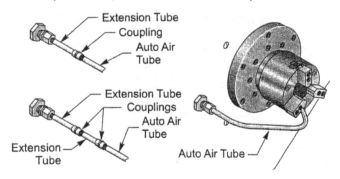

Haas CNC Lathe
Parts Catcher

The parts catcher is an automatic part retrieval system designed to work with bar feed applications. The parts catcher, is commanded using M36 to activate and M37 to deactivate. The parts catcher rotates to catch finished parts and directs them into the bin mounted on the front door. The parts catcher must be properly aligned before operation.

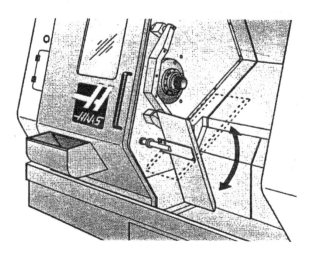

Power on the machine. In MDI mode, activate the parts catcher (M36).

The operator door must be closed when actuating the part catcher.

Loosen the screw in the shaft collar on the outer parts catcher shaft.

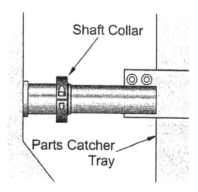

Continued on next page.

Haas CNC Lathe
Parts Catcher

Slide parts catcher tray into the shaft far enough to catch the part and clear the chuck. Rotate the tray to open the sliding cover of the parts collector mounted in the door and tighten the shaft collar on the part catcher shaft.

Check the Z-axis, X-axis, tool and turret position during part catcher actuation to avoid possible collisions during operation.

When programming the part catcher in a program you must use a G04 code between M53 and M63 to pause the catcher pan in the open position long enough to cut off the part and allow it to fall into the collector.

Large chuck jaws may interfere with the operation of the parts catcher. Be sure to check the clearances before operating the parts catcher.

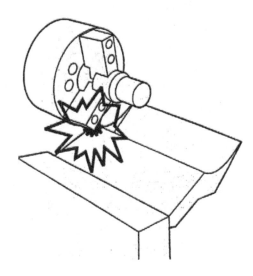

CAD-RESOURCES.COM
Computer Aided Design Classes

De Anza College
CDI
Department Head
& Instructor
of
Pro/ENGINEER
Pro/SHEETMETAL
Pro/SURFACE
&
Unigraphics NX

Louis Gary Lamit:
Is currently a full time instructor and department head at De Anza College in Cupertino, California, where he teaches Pro/ENGINEER, Pro/SURFACE, Pro/SHEETMETAL, Pro/NC, and Expert Machinist, and Unigraphics NX. He also owns and operates his own consulting firm, and has been involved with advertising, and patent illustration.

He is the author of over 30 textbooks, workbooks, tutorials, and handbooks. Mr. Lamit received a BS degree from Western Michigan University in 1970 and did Masters' work at Wayne State University and Michigan State University. He has also done graduate work at the University of California at Berkeley and holds an NC programming certificate from Boeing Aircraft. Since leaving industry, he has taught at several Colleges and Universities before joining the staff at De Anza College, Cupertino, California.

Allowances & Fits

Factors Affecting Selection of Fits:

Many factors, such as length of engagement, bearing load, speed, lubrication, temperature, humidity, and materials, must be taken into consideration in the selection of fits for a particular application, and modifications in the ANSI standards recommendations may be required to satisfy extreme conditions. Subsequent adjustments may also be found desirable as a result of experience in a particular application to suite critical functional requirements or to permit optimum manufacturing economy.

Note: For ball and roller bearings such as SKF, Fafnir, Timken, or Torrington, always use the bearing manufactures recommended fits.

If the manufactures recommended fits are not available see Machinery's Handbook, 25th edition or later.

Note: For quick and easy press fits or running fits for nominal diameters make use of the over and under reamers that are in most machine shop tool cribs.

Allowances & Fits: Starrett

The L. S. Starrett Company "the world's greatest toolmakers" have been in business since 1880 and in the back of their catalog they have their chart for "Allowances for Fits" Their chart and formulas are reproduced below.

Formulas for Determining Allowances

Class	High Limit	Low Limit
A	$+\sqrt{D \times 0.0006}$	$-\sqrt{D \times 0.0003}$
B	$+\sqrt{D \times 0.0008}$	$-\sqrt{D \times 0.0004}$
P	$-\sqrt{D \times 0.0002}$	$-\sqrt{D \times 0.0006}$
X	$-\sqrt{D \times 0.00125}$	$-\sqrt{D \times 0.0025}$
Y	$-\sqrt{D \times 0.001}$	$-\sqrt{D \times 0.0018}$
Z	$-\sqrt{D \times 0.0005}$	$-\sqrt{D \times 0.001}$

Allowances & Fits: Starrett
Tolerances in Standard Holes

Tolerance is provided for holes, which ordinary standard reamers can produce, in two grades, Class A and B, the selection of which is a question for the users decision and dependent upon the quality of the work required; some prefer to use Class A as working limits and Class B as inspection limits.

Hole Sizes Up To 2-Inch Diameter

Class	Nominal Dia.	Up To 1/2"	9/16 to 1"	1-1/16 to 2"
A	High Limit	+0.00025	+0.0005	+0.00075
	Low Limit	−0.00025	−0.00025	−0.00025
	Tolerance	0.0005	0.00075	0.0010
B	High Limit	+0.0005	+0.00075	+0.0010
	Low Limit	−0.0005	−0.0005	−0.0005
	Tolerance	0.0010	0.00125	0.0015

Hole Sizes Up To 5-Inch Diameter

Class	Nominal Dia.	2-1/16 To 3"	3-1/16 to 4"	4-1/16 to 5"
A	High Limit	+0.0010	+0.0010	+0.0010
	Low Limit	−0.0005	−0.0005	−0.0005
	Tolerance	0.0015	0.0015	0.0015
B	High Limit	+0.00125	+0.0015	+0.00175
	Low Limit	−0.00075	−0.00075	−0.00075
	Tolerance	0.0020	0.00225	0.0025

ATC Building, CDI: CAD Room 301 & 313

Allowances & Fits: Starrett
Allowances for Forced Fits (Shaft)

Hole Sizes Up To 2-Inch Diameter

Class		Up To 1/2"	9/16 to 1"	1-1/16 to 2"
	Nominal Dia.			
F	High Limit	+0.0010	+0.0020	+0.0040
	Low Limit	+0.0005	+0.0015	+0.0030
	Tolerance	0.0005	0.0005	0.0010

Hole Sizes Up To 5-Inch Diameter

Class		2-1/16 To 3"	3-1/16 to 4"	4-1/16 to 5"
	Nominal Dia.			
F	High Limit	+0.0060	+0.0080	+0.0100
	Low Limit	+0.0045	+0.0060	+0.0080
	Tolerance	0.0015	0.0020	0.0020

Allowances for Driving Fits (Shaft)

Hole Sizes Up To 2-Inch Diameter

Class		Up To 1/2"	9/16 to 1"	1-1/16 to 2"
	Nominal Dia.			
D	High Limit	+0.0005	+0.0010	+0.0015
	Low Limit	+0.00025	+0.00075	+0.0010
	Tolerance	0.00025	0.00025	0.0005

Hole Sizes Up To 5-Inch Diameter

Class		2-1/16 To 3"	3-1/16 to 4"	4-1/16 to 5"
	Nominal Dia.			
D	High Limit	+0.0025	+0.0030	+0.0035
	Low Limit	+0.0015	+0.0020	+0.0025
	Tolerance	0.0010	0.0010	0.0010

Allowances for Push Fits (Shaft)

Hole Sizes Up To 2-Inch Diameter

Class		Up To 1/2"	9/16 to 1"	1-1/16 to 2"
	Nominal Dia.			
P	High Limit	-0.00025	-0.00025	-0.00025
	Low Limit	-0.00075	-0.00075	-0.00075
	Tolerance	0.0005	0.0005	0.0005

Hole Sizes Up To 5-Inch Diameter

Class		2-1/16 To 3"	3-1/16 to 4"	4-1/16 to 5"
	Nominal Dia.			
P	High Limit	-0.0005	-0.0005	-0.0005
	Low Limit	-0.0010	-0.0010	-0.0010
	Tolerance	0.0005	0.0005	0.0005

Allowances & Fits: Starrett
Allowances for Running Fits (Shaft)

Running fits, which are the most commonly required, are divided into three grades: Class X, for engine and other work where easy fits are wanted: Class Y, for high speeds and good average machine work: Class Z, for fine tool work.

Hole Sizes Up To 2-Inch Diameter

Class	Nominal Dia.	Up To 1/2"	9/16 to 1"	1-1/16 to 2"
X	High Limit	-0.0010	-0.00125	-0.00175
	Low Limit	-0.0020	-0.00275	-0.0035
	Tolerance	0.0010	0.0015	0.00175
Y	High Limit	-0.00075	-0.0010	-0.00125
	Low Limit	-0.00125	-0.0020	-0.0025
	Tolerance	0.0005	0.0010	0.00125
Z	High Limit	-0.0005	-0.00075	-0.00075
	Low Limit	-0.00075	-0.00125	-0.0015
	Tolerance	0.00025	0.0005	0.00075

Hole Sizes Up To 5-Inch Diameter

Class	Nominal Dia.	2-1/16 To 3"	3-1/16 to 4"	4-1/16 to 5"
X	High Limit	-0.0020	-0.0025	-0.0030
	Low Limit	-0.00425	-0.0050	-0.00575
	Tolerance	0.00225	0.0025	0.00275
Y	High Limit	-0.0015	-0.0020	-0.00225
	Low Limit	-0.0030	-0.0035	-0.0040
	Tolerance	0.0015	0.0015	0.00175
Y	High Limit	-0.0010	-0.0010	-0.00125
	Low Limit	-0.0020	-0.00225	-0.0025
	Tolerance	0.0010	0.00125	0.00125

Allowances & Fits: ASA B4a-1925

The American Society of Mechanical Engineers (ASME) formulas and comments, published below were extracted from the ASA B4a-1925 standard published in the ASME Handbook, Engineering Tables (1956 edition). These standards have been in common use for many decades and are now replaced by USAS B4.1-1967

From	Up To & Including	Mean Diameter (d)	Cube Root of Mean Diameter (d)
0	3/16	.125	0.50000000
3/15	5/16	.250	0.62996052
5/16	7/16	.375	0.72112478
7/16	9/16	.500	0.79370052
9/16	11/16	.625	0.85498797
11/16	13/16	.750	0.90856029
13/16	15/16	.875	0.95646559
15/16	1-1/16	1.000	1.00000000
1-1/16	1-3/16	1.125	1.04004191
1-3/16	1-3/8	1.250	1.07721735
1-3/8	1-5/8	1.500	1.14471424
1-5/8	1-7/8	1.750	1.20507113
1-7/8	2-1/8	2.000	1.25992105
2-1/8	2-3/8	2.250	1.31037070
2-3/8	2-3/4	2.500	1.35720881
2-3/4	3-1/4	3.000	1.44224957
3-1/4	3-3/4	3.500	1.51829449
3-3/4	4-1/4	4.000	1.58740105
4-1/4	4-3/4	4.500	1.65096362
4-3/4	5-1/2	5.000	1.70997595
5-1/2	6-1/2	6.000	1.81712059
6-1/2	7-1/2	7.000	1.91293118
7-1/2	8-1/2	8.000	2.00000000

Allowances & Fits: ASA B4a-1925

Class 1, Large Allowance, Interchangeable
This fit provides for considerable freedom and embraces certain fits where accuracy is not essential.

When d = Mean Diameter   (See Chart for Mean Diameter)

Hole Tolerance (+) = $0.0025 \times \sqrt[3]{d}$

Shaft Tolerance (-) = $0.0025 \times \sqrt[3]{d}$

Allowance = $0.0025 \times \sqrt[3]{d^2}$

Class 2, Liberal Allowance, Interchangeable
For running fits with speeds of 600 rpm or over, and journal pressures of 600 lbs. per square inch or over.

When d = Mean Diameter   (See Chart for Mean Diameter)

Hole Tolerance (+) = $0.0013 \times \sqrt[3]{d}$

Shaft Tolerance (-) = $0.0013 \times \sqrt[3]{d}$

Allowance = $0.0014 \times \sqrt[3]{d^2}$

Class 3, Medium Allowance, Interchangeable
For running fits under 600 rpm and with journal pressures less than 600 lbs. Per square inch; also for sliding fits, and the more accurate machine-tool and automotive parts.

When d = Mean Diameter   (See Chart for Mean Diameter)

Hole Tolerance (+) = $0.0008 \times \sqrt[3]{d}$

Shaft Tolerance (-) = $0.0008 \times \sqrt[3]{d}$

Allowance = $0.0009 \times \sqrt[3]{d^2}$

Allowances & Fits: ASA B4a-1925

Class 4, Zero Allowance, Interchangeable
This is the closest fit which can be assembled by hand and necessitates work of considerable precision. It should be used where no perceptible shake is permissible and where moving parts are not intended to move freely under load.

When d = Mean Diameter (See Chart for Mean Diameter)

Hole Tolerance (+) = $0.0006 \times \sqrt[3]{d}$

Shaft Tolerance (-) = $0.0004 \times \sqrt[3]{d}$

Allowance = 0.0000

Class 5, Zero to Negative Allowance, Selective Assy.
This is also known as a "tunking fit" and is practically metal-to-metal. Assembly is usually selective and not interchangeable. The average interference of metal is the desired condition and must be obtained by selective assembly that is, by mating large shafts in large holes and small shafts in small holes.

When d = Mean Diameter (See Chart for Mean Diameter)

Hole Tolerance (+) = $0.0006 \times \sqrt[3]{d}$

Shaft Tolerance (-) = $0.0004 \times \sqrt[3]{d}$

Average interference of metal = 0.0000

Class 6, Slight Negative Allowance, Selective Assy.
Light pressure is required to assemble these fits and the parts are more or less permanently assembled, such as the fixed ends of studs for gears pulleys, rocker arms, etc. These fits are used for drive fits in thin sections or extremely long fits in other sections, and also for shrink fits on very light sections. Used in automotive, ordnance, and general machine manufacturing. The average interference of metal is the desired condition and must be obtained by selective assembly that is, by mating large shafts in large holes and small shafts in small holes.

When d = Mean Diameter (See Chart for Mean Diameter)

Hole Tolerance (+) = $0.0006 \times \sqrt[3]{d}$

Shaft Tolerance (-) = $0.0006 \times \sqrt[3]{d}$

Average interference of metal = $0.00025 \times d$

Allowances & Fits: ASA B4a-1925

Class 7, Negative Allowance, Selective Assy.
Considerable pressure is required to assemble these fits and the parts are considered permanently assembled. These fits are used in fastening locomotive wheels, car wheels, armatures of dynamos and motors, and crank disks to their axles or shafts. They are also used for shrink fits on medium sections or long fits. These fits are the tightest which are recommended for cast-iron holes or external members as they stress cast-iron to its elastic limit. The average interference of metal is the desired condition and must be obtained by selective assembly that is, by mating large shafts in large holes and small shafts in small holes. For hole and shaft tolerances, the same formulas were used for sizes larger than 8 inch, although there was no data available for these diameters.

When d = Mean Diameter   (See Chart for Mean Diameter)

Hole Tolerance (+) = $0.0006 \times \sqrt[3]{d}$

Shaft Tolerance (-) = $0.0006 \times \sqrt[3]{d}$

Average interference of metal = $0.0005 \times d$

Class 8, Considerable Negative Allowance, Selective Assy.
These fits are used for steel holes where the metal can be stressed to its elastic limit. These fits cause excessive stress for cast-iron holes. Shrink fits are used where heavy force fits are impractical, as on locomotive wheel tires, heavy crank disks of large engines, etc. The average interference of metal is the desired condition and must be obtained by selective assembly that is, by mating large shafts in large holes and small shafts in small holes. For hole and shaft tolerances, the same formulas were used for sizes larger than 8 inch, although there was no data available for these diameters.

When d = Mean Diameter   (See Chart for Mean Diameter)

Hole Tolerance (+) = $0.0006 \times \sqrt[3]{d}$

Shaft Tolerance (-) = $0.0006 \times \sqrt[3]{d}$

Average interference of metal = $0.001 \times d$

Allowances & Fits:
ANSI B4.1-1967, Class Of Fits

The American Society of Mechanical Engineers (ASME) tables for allowances and fits are to extensive to include in this booklet and their new standards do not include any formulas. Only the class of fits is published below. The Department of Defense (DOD), and other U.S. Government agencies now require this standard.

For a more complete description of the USAS B4.1-1967 (ANSA B4.1-1967) standard see Machinery's Handbook, twenty-fifth edition or later. You can purchase a copy of this standard from, www.document-center.com.

Sliding Fits

RC 1, Close sliding fits are intended for accurate location of parts which must assemble without perceptible play.

RC 2, Sliding fits are intended for accurate location but with grater maximum clearance than class RC 1. parts made to this fit move and turn easily but are not intended to run freely, and in larger size may seize with small temperature changes.

Running Fits

RC 3, Precision running fits are about the closest fits which can be expected to run freely, and are intended for precision work at slow speeds and light journal pressures, but are not suitable where appreciable temperature differences are likely to be encountered.

RC 4, Close running fits are intended chiefly for running fits on accurate machinery with moderate surface speeds and journal pressures, where accurate location and minimum play is desired.

RC 5 & RC 6, Medium running fits are intended for higher running speeds, or heavy journal pressures, or both.

RC 7, Free running fits are intended for use where accuracy is not essential, or where large temperature variations are likely to be encountered, or under both of these conditions.

RC 8 & RC 9, Loose running fits are intended for use where wide commercial tolerances may be necessary, together with an allowance, on the external member.

Allowances & Fits:
ANSI B4.1-1967, Class Of Fits

Locational Fits

LC 1 through LC 11,
Locational clearance fits
are intended for parts which are normally stationary, but which can be freely assembled or disassembled. They run from snug fits for parts requiring accuracy of location, through the medium clearance fits for parts such as ball, race and housing, to the looser fastener fits where freedom of assembly is of prime importance.

LT 1 through LT 6,
Locational transition fits are compromise between clearance and interference fits, for application where accuracy of location is important, but either a small amount of clearance or interference is permissible.

LN 1 through LN 3,
Locational interference fits are used where accuracy of location is of prime importance and for parts requiring rigidity and alignment with no special requirements for bore pressure. such fits are not intended for parts designed to transmit frictional loads from one part to another by virtue of the tightness of fit, as these conditions are covered by force fits.

Drive Fits

FN 1, Light drive fits
are those requiring light assembly pressures and produce more or less permanent assemblies. They are suitable for thin sections or long fits, or in cast-iron external members.

FN 2, Medium drive fits
are suitable for ordinary steel parts or for shrink fits on light sections. They are about the tightest fits that can be used with high-grade cast-iron external members.

FN 3, Heavy drive fits
are suitable for heavier steel parts or for shrink fits in medium sections.

FN 4 & FN 5, Force fits
are suitable for parts which can be highly stressed or for shrink fits where the heavy pressing forces required are impractical.

Drilling & Reaming

Drills

High-Helix Drills – sometimes called fast spiral drills, are designed to remove chips from deep holes. The large rake angle makes these drills suitable for soft metals such as aluminum and mild steel.

Low-Helix Drills – sometimes called slow spiral drills, are more rigid than standard helix drills and can stand more torque. Like straight-fluted drills, they are less likely to grab when emerging from a hole because of the small rake angle.

Straight-fluted drills – are used for drilling brass and other soft materials because the zero rake angle eliminates the tendency for the drill to grab on breakthrough. For the same reason they are used on thin materials.

Standard-fluted drills – this twist drill is by far the most common type of drill used today. These are made with two or more flutes and cutting lips and in many varieties of tip design and lengths.

Reamers

Straight-flute chucking reamers – have shanks that are slightly smaller than the reamer diameter, except for wire gauge size number 61 to 80, which have identical shank and reamer diameters. Straight-flute reamers are perfect for most general purpose reaming job.

Spiral-flute chucking reamers – produce smother chatter free cuts than straight-flute reamers. They easily guide work, and have a right-hand spiral and cut. Use them for hard-to-ream materials and to help eliminate chips in blind holes.

Oversize & Undersize chucking reamers – making a slightly looser or tighter hole fit is easy when you use these straight- shank, straight-flute reamers, oversize reamers are +.001" and undersize reamers are -.001" from common fractional sizes, all reamers have a 45° chamfer angle and a cutting diameter tolerance of +.0002 / -.0000 inch.

Drilled Hole Tolerances

Hole Size	Tolerances
.0135 to .1850 =	+.003 -.002
.1875 to .2460 =	+.004 -.002
.2500 to .7500 =	+.005 -.002
.7656 to 1.000 =	+.007 -.003

Drilling Feed Table

Drill Size	Feed Per Rev
Under 1/8"	.001 to .002
1/8" to 1/4"	.002 to .004
1/4" to 1/2"	.004 to .007
1/2" to 1"	.007 to .015
Over 1"	.015 to .0025

Reaming Stock Allowance

Reamer Size	Hole Undersize Allowance
1/32" to 1/8"	.003 to .006
1/8" to 1/4"	.005 to .009
1/4" to 3/8"	.007 to .012
3/8" to 1/2"	.010 to .015
1/2" to 3/4"	1/64" to 1/32"
3/4" to 1.0"	1/32"

Center Drill Dimensions
Standard 60° Center Drill

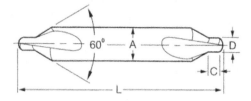

Size	Body Dia (A)	Drill Dia (D)	Body Lg (C)	Overall Lg (L)
00	1/8	.0625	0.030	1-1/8
0	1/8	1/32	0.038	1-1/8
1	1/8	3/64	3/64	1-1/4
2	3/16	5/64	5/64	1-7/8
3	1/4	7/64	7/64	2
4	5/16	1/8	1/8	2-1/8
5	7/16	3/16	3/16	2-3/4
6	1/2	7/32	7/32	3
7	5/8	1/4	1/4	3-1/4
8	3/4	5/16	5/16	3-1/2

Tolerance:
A Diameter = Plus 0.0000, Minus 0.0020
D Diameter = Plus 0.0030, Minus 0.0000
C Length for sizes 00 to 2 = Plus .008, minus .008
C Length for sizes 3 to 8 = Plus 1/64, minus 1/64
L Overall Length = Plus 1/32, Minus 1/16

Center Drill, Drill Point Allowance

"Z" Depth & Drill Size			Hole			
#1	#2	#3	Dia.			
.058			.050	To program Z-axis depth when center drilling. Locate required diameter "D" and the "Z" value for the size center drill being used.		
.067			.060			
.075			.070			
.084	.094		.080			
.093	.103		.090			
.101	.111		.100			
.110	.120	.130	.110	"Z" Depth & Drill Size		
.119	.129	.138	.120	#4	#5	#6
	.137	.147	.130	.152		
	.146	.156	.140	.161		
	.155	.164	.150	.169		
	.163	.173	.160	.178		

Center Drill Dimensions
Standard 60° Center Drill
Center Drill, Drill Point Allowance

"Z" Depth & Drill Size			Hole Dia.	"Z" Depth & Drill Size		
#1	#2	#3		#4	#5	#6
	.172	.182	.170	.187		
	.181	.190	.180	.195		
		.199	.190	.204	.224	
		.208	.200	.213	.232	
		.216	.210	.221	.241	
		.225	.220	.230	.250	
		.234	.230	.239	.258	.268
		.242	.240	.247	.267	.277
		.251	.250	.256	.276	.286
			.260	.265	.284	.295
			.270	.273	.293	.303
			.280	.282	.302	.312
			.290	.291	.310	.320
			.300	.299	.319	.329
			.310	.308	.328	.338
			.320		.336	.346
			.330		.345	.355
			.340		.354	.364
			.350		.362	.372
			.360		.371	.381
			.370		.380	.390
			.380		.388	.398
			.390		.397	.407
			.400		.405	.416
			.410		.414	.424
			.420		.423	.433
			.430		.431	.442
			.440			.450
			.450			.459
			.460			.468
			.470			.477
			.480			.485
			.490			.494
			.500			.502

To program Z-axis depth when center drilling. Locate required diameter "D" and the "Z" value for the size center drill being used.

"Z" DEPTH

"D" DIAMETER

Drill Point Depth
Countersink Allowance

Drill Point Angle	Allowance
60°	0.866 x Dia. = Point Depth
82°	0.575 x Dia. = Point Depth
90°	0.500 x Dia. = Point Depth
100°	0.419 x Dia. = Point Depth
118°	0.300 x Dia. = Point Depth
120°	0.288 x Dia. = Point Depth
135°	0.207 x Dia. = Point Depth

To calculate drill depth for a chamfer diameter, or point depth for a required drilling depth.

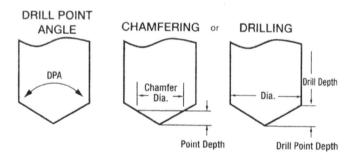

Spotting Drills
Center Drill

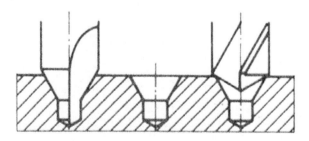

Center hole produced with combined drill and countersink. The small pilot diameter requires a slower feed rate for the centering operation than the twist drill for drilling. A further loss in time results from the necessary centering depth. The following twist drill only starts cutting at the outer lips, with the consequence of poor alignment and reduced tool life.

Spotting Drills

Centering Drill

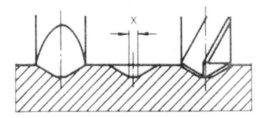

Center hole, produced by centering drill provided with too thick a web. The chisel edge width determines the dimensions "X." Therefore the follow-on tool must be designed with a larger web thickness or chisel edge width. The tool would otherwise not be centered and the purpose of centering would be totally missed.

NC-Spotting Drill

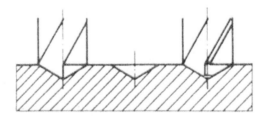

Center holes, produced with NC-spotting drills that have extremely short chisel edges, forms an almost pointed centering cone offering a "tight" guidance for the follow-on tool. A prerequisite, however is that an identical point angle of NC-spotting drill and twist drill has been provided.

90-degree NC-Spotting Drill

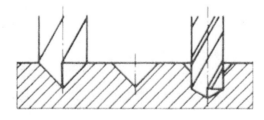

If the diameter of the NC-spotting drill is larger than that of the twist drill used for the subsequent machining cycle. Centering and chamfering are made in only one operation. NC-spotting drills with a point angle of 90-degrees is most advantageous for this purpose.

Drill Point Angles

Fig 1 & 2: Average class of work (mild steel). 118° included angle, 12° to 15° lip clearance.

Fig 3: Alloy steels, monel, stainless steel, heat-treated steels, drop forgings. Brinell hardness #240. 125° included angle, 10° to 12° lip clearance.

Fig 4: Soft and medium cast iron, aluminum, marble, slate, plastics, wood, hard rubber, bakelite. 90° to 130° included angle, 12° lip clearance. Flat cutting lip for marble.

Fig (none): Copper, soft and medium hard brass. 100° to 118° included angle, 12° to 15° lip clearance.

Fig 5: Magnesium alloys. 60° to 118° included angle, 15° lip clearance. Slightly flat face cutting lips reducing rake angle to 5°

Fig 6: Wood, rubber, bakelite, fibre, aluminum die castings, plastics. 60° included angle, 12° to 15° lip clearance.

Fig 7: Steel 7% manganese, tough alloy steels, armor plate and hard materials. 150° included angle, 7° to 10° lip clearance. Slight flat on face of cutting lips.

Fig 8: Brass, soft bronze. 118° included angle, 12° to 15° lip clearance. Slight flat on face of cutting lips.

Fig 9: Crankshafts, deep holes in soft steel, hard steel, cast iron, nickel and manganese alloys. 118° included angle, 9° lip clearance.

Fig 10: Thin sheet metal, copper, fibre, plastics and wood. -5° to +12° lip angles, for drills over 1/4" Dia. make angle of bit point to suite work.

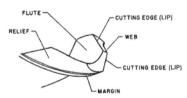

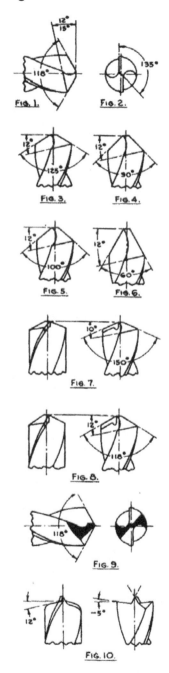

Decimal Equivalents

Drill Size	Dec.	Drill Size	Dec.	Drill Size	Dec.
0.1mm	.0039	54	.0550	26	.1470
0.2mm	.0079	1.5mm	.0591	25	.1495
0.3mm	.0118	53	.0595	24	.1520
80	.0135	1/16	.0625	23	.1540
79	.0145	52	.0635	5/32	.1562
1/64	.0156	51	.0670	22	.1570
0.4mm	.0157	50	.0700	4mm	.1575
78	.0160	49	.0730	21	.1590
77	.0180	48	.0760	20	.1610
0.5mm	.0197	5/64	.0781	15	.1660
76	.0200	47	.0785	18	.1695
75	.0210	2mm	.0787	11/64	.1719
74	.0225	46	.0810	17	.1730
0.6mm	.0236	45	.0820	16	.1770
73	.0240	44	.0860	4.5mm	.1772
72	.0250	43	.0890	15	.1800
71	.0260	42	.0935	14	.1820
0.7mm	.0276	3/32	.0938	13	.1850
70	.0280	41	.0960	3/16	.1875
69	.0292	40	.0980	12	.1890
68	.0310	2.5mm	.0984	11	.1910
1/32	.0312	39	.0995	10	.1935
0.8mm	.0315	38	.1015	9	.1960
67	.0320	37	.1040	5mm	.1969
66	.0330	36	.1065	8	.1990
65	.0350	7/64	.1094	7	.2010
0.9mm	.0354	35	.1100	13/64	.2031
64	.0360	34	.1110	6	.2040
63	.0370	33	.1130	5	.2055
62	.0380	32	.1160	4	.2090
61	.0390	3mm	.1181	3	.2130
1mm	.0394	31	.1200	5.5mm	.2165
60	.0400	1/8	.1250	7/32	.2188
59	.0410	30	.1285	2	.2210
58	.0420	29	.1360	1	.2280
57	.0430	3.5mm	.1378	A	.2340
56	.0465	28	.1405	15/64	.2344
3/64	.0469	9/64	.1406	6mm	.2362
55	.0520	27	.1440	B	.2380

Decimal Equivalents

Drill Size	Dec.	Drill Size	Dec.	Drill Size	Dec.
C	.2420	X	.3970	23/32	.7188
D	.2460	Y	.4040	18.5mm	.7283
E	.2500	13/32	.4062	47/64	.7344
1/4	.2500	Z	.4130	19mm	.7480
6.5mm	.2550	10.5mm	.4134	3/4	.7500
F	.2570	27/64	.4219	49/64	.7656
G	.2610	11mm	.4331	19.5mm	.7677
17/64	.2656	7/16	.4375	25/32	.7812
H	.2660	11.5mm	.4528	20mm	.7874
I	.2720	29/64	.4531	51/64	.7969
7mm	.2756	15/32	.4688	20.5mm	.8071
J	.2770	12mm	.4724	13/16	.8125
K	.2810	31/64	.4844	21mm	.8268
9/32	.2812	12.5mm	.4821	53/64	.8281
L	.2900	1/2	.5000	27/32	.8438
M	.2950	13mm	.5118	21.5mm	.8465
7.5mm	.2953	33/64	.5156	55/64	.8594
19/64	.2969	17/34	.5313	22mm	.8661
N	.3020	13.5mm	.5315	7/8	.8750
5/16	.3125.	35/64	.5469	22.5mm	.8858
8mm	.3150	14mm	.5512	57/64	.8906
O	.3160	9/16	.5625	23mm	.9065
P	.3230	14.5mm	.5709	29/32	.9062
21/64	.3281	37/64	.5781	59/64	.9219
Q	.3320	15mm	.5906	23.5mm	.9252
8.5mm	.3346	19/32	.5938	15/16	.9375
R	.3390	39/64	.6094	24mm	.9449
11/32	.3438	15.5mm	.6102	61/64	.9531
S	.3480	5/8	.6250	24.5mm	.9646
9mm	.3543	16mm	.6299	31/32	.9688
T	.3580	41/64	.6406	25mm	.9843
23/64	.3594	16.5mm	.6496	63/64	.9844
U	.3680	21/32	.6562	1	1.000
9.5mm	.3740	17mm	.6693		
3/8	.3750	43/64	.6719		
V	.3770	11/16	.6875		
W	.3860	17.5mm	.6890		
25/64	.3906	45/64	.7031		
10mm	.3937	18mm	.7067		

Countersinks & Counterbores

Countersinks:

Screw hole countersinks listed hear are based on the flat head screw's maximum head diameter with sharp edge.

Flat head screws actually have three head diameters, and three head angles as listed below. Flat head screws also have at least seven or more head styles.

Flat Head Screw, Head Diameters:
Edge sharp, maximum diameter
(see tables)

Edge sharp, minimum diameter

Edge rounded or flat, minimum diameter

Flat Head Screw Styles & Angles:
Aircraft, 100°

Standard, 80° to 82°

Standard, Oval, 80° to 82°

Standard, Undercut, 80° to 82°

Standard, Machinist, 80° to 82°
(Socked Head)

Metric, 90° to 92°

Metric, Machinist, 90° to 92°
(Socked Head)

Counterbores:

Screw hole counterbores listed hear are based on commercially available piloted-counterbores that are commonly used on manual mills such as the Bridgeport.

This will allow you to produce the same counterbore size holes on a CNC as would be produced on a Bridgeport.

Do not use piloted-counterbores on CNC miles or lathes.

Countersinks, Blunt Ended

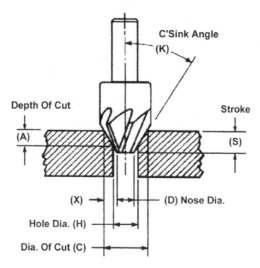

Angle	Angle (K)	Ratio (R) = 2(X / A)
60°	30°	1.15
82°	41°	1.74
90°	45°	2.00
100°	50°	2.38
110°	55°	2.86
120°	60°	3.46

Angle (K) Tangent = X / A

Calculations

C = H + RA
A = (C − H) / R
S = (C − D) / R
X = (C − H) / 2

Note:
For any depth the countersink diameter expands in a direct ratio. As the countersink travels along the Z-axis into the hole the holes radius expands by (X) in a direct relationship to angle (K); forming an angle with the tangent of "X / A". Since the angle expands on both sides of the drilled hole; you must use 2 times the tangent of the angle for your ratio (R).

Example: .500" hole diameter, countersink to .875" cut diameter with 41° angle.
A = (.875 − .500) / 1.74 = .216"
S = (.875 − .203) / .203 = .386"

Counterbores
Socket Head Cap Screw, Inch

Screw Size	Through Hole Diameter	Counter Bore Diameter	Counter Bore Depth (Flush Head)
#0	.060	.100	.060
"	.064	.104	"
"	.070	.110	"
#1	.073	.120	.073
"	.078	.128	"
"	.081	.140	"
#2	.086	.144	.086
"	.094	.147	"
"	.096	.161	"
#3	.099	.166	.099
"	.104	.177	"
"	.110	.185	"
#4	.112	.186	.112
"	.127	.202	"
"	.143	.218	"
#5	.125	.205	.125
"	.140	.221	"
"	.156	.237	"
#6	.138	.228	.138
"	.153	.244	"
"	.169	.260	"
#8	.164	.272	.164
"	.179	.288	"
"	.195	.304	"
#10	.190	.314	.190
"	.205	.330	"
"	.221	.346	"
1/4"	.250	.382	.250
"	.265	.398	"
"	.281	.414	"
5/16"	.312	.475	.312
"	.328	.491	"
"	.343	.507	"
3/8"	.375	.573	.375
"	.390	.588	"
"	.406	.604	"

Counterbores
Socket Head Cap Screw, Inch

Screw Size	Through Hole Diameter	Counter Bore Diameter	Counter Bore Depth (Flush Head)
7/16"	.437	.663	.437
"	.452	.679	"
"	.468	.695	"
1/2"	.500	.757	.500
"	.515	.773	"
"	.531	.789	"
9/16"	.562	.859	.562
"	.875	.890	"
"	.906	.912	"
5/8"	.625	.945	.625
"	.640	.961	"
"	.656	.977	"
3/4"	.750	1.133	.750
"	.765	1.149	"
"	.781	1.165	"
7/8"	.875	1.322	.875
"	.891	1.338	"
"	.906	1.354	"
1"	1.000	1.510	1.000
"	1.015	1.526	"
"	1.031	1.542	"

Paul Klingman, SolidWorks Instructor

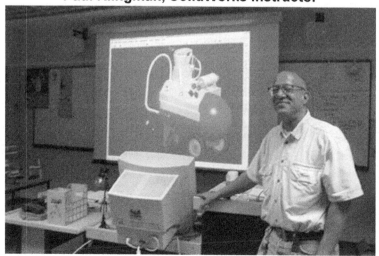

Countersinks
Flat Head Screws, Inch

Screw Size	Through Hole Dia	82° Csk Max Dia. (Sharp Edge)	100° Csk Max Dia. (Sharp Edge)
#0	.060	.119	.119
"	.064	"	"
"	.070	"	"
#1	.073	.146	.146
"	.078	"	"
"	.081	"	"
#2	.086	.172	.172
"	.094	"	"
"	.096	"	"
#3	.099	.199	.199
"	.104	"	"
"	.110	"	"
#4	.112	.225	.225
"	.127	"	"
"	.143	"	"
#5	.125	.252	.252
"	.140	"	"
"	.156	"	"
#6	.138	.279	.279
"	.153	"	"
"	169	"	"
#8	.164	.332	.332
"	.179	"	"
"	.195	"	"
#10	.190	.385	.385
"	.205	"	"
"	.221	"	"
1/4"	.250	.507	.507
"	.265	"	"
"	.281	"	"
5/16"	.312	.635	.635
"	.328	"	"
"	.343	"	"
3/8"	.375	.762	.762
"	.390	"	"
"	.406	"	"

Countersinks
Flat Head Screws, Inch

Screw Size	Through Hole Dia.	82° Countersink Max Dia. (Sharp Edge)	100° Countersink Max Dia. (Sharp Edge)
7/16"	.437	.812	.812
"	.452	"	"
"	.468	"	"
1/2"	.500	.875	.875
"	.515	"	"
"	.531	"	"
5/8"	.625	1.125	1.125
"	.640	"	"
"	.656	"	"
3/4"	.750	1.375	1.375
"	.765	"	"
"	.781	"	"

SolidWorks Class

Counterbores
Socket Head Cap Screws, Metric

Screw Size	Through Hole Diameter	Counter Bore Diameter	Counter Bore Depth (Flush Head)
M 1.6	1.8	3.5	1.6
"	1.95	"	"
M 2	2.2	4.4	2.0
"	2.4	"	"
M 2.5	2.7	5.4	2.5
"	3.0	"	"
M 3	3.4	6.5	3.0
"	3.7	"	"
M 4	4.4	8.25	4.0
"	4.8	"	"
M 5	5.4	9.75	5.0
"	5.8	"	"
M 6	6.4	11.2	6.0
"	6.8	"	"
M 8	8.4	14.5	8.0
"	8.8	"	"
M 10	10.5	17.5	10.0
"	10.8	"	"
M 12	12.5	19.5	12.0
"	13.0	"	"
M 14	14.5	22.5	14.0
"	15.0	"	"
M 16	16.5	25.5	16.0
"	17.0	"	"
M 20	20.5	31.5	20.0
"	21.0	"	"
M 24	24.5	37.5	24.0
"	25.0	"	"
M 30	31.0	47.5	30.0
"	31.5	"	"
M 36	37.0	56.5	36.0
"	37.5	"	"
M 42	43.0	66.0	42.0
"	44.0	"	"
M 48	49.0	75.0	48.0
"	50.0	"	"

Countersinks
Flat Head Screws, Metric

Screw Size	Through Hole Diameter	90° Countersink Max Diameter (Sharp Edge)
M 1.6	1.8	
"	1.95	
M 2	2.2	4.0
"	2.4	"
M 2.5	2.7	5.0
"	3.0	"
M 3	3.4	6.72
"	3.7	"
M 4	4.4	8.96
"	4.8	"
M5	5.4	11.2
"	5.8	"
M 6	6.4	13.44
"	6.8	"
M 8	8.4	17.92
"	8.8	"
M 10	10.5	22.4
"	10.8	"
M 12	12.5	26.88
"	13.0	"
M 14	14.5	30.24
"	15.0	"
M 16	16.5	33.6
"	17.0	"
M 20	20.5	40.32
"	21.0	"

Tapping & Threading

Roll Form Taps:
Thread forming taps cold form rather than cut the threads. Advantages include no chips to dispose of, stronger higher quality threads, increased tapping speeds, longer tap life and reduced tap breakage. Recommended for CNC tapping of a wide variety of ductile materials up to 28Rc hardness.

Shear Taps:
Used in steels and stainless steels up to 35Rc hardness, such as 1010, 1018, 1046, 302, 303, 304, 310, or 316.

Spiral Flute Taps:
Regular spiral fluted (35-degree helix) taps for blind-hole applications to depths up to two-times the tap diameter in a variety of materials including free machining steels, alloyed or unalloyed steels and ductile cast irons.

Fast Spiral Fluted Taps:
(45-degree helix) for blind or bottoming hole applications to depths up to three-times the tap diameter in a variety of stringy materials. The higher helix angle increases chip removal efficiency.

Standard Taps:
Use these taps for general machine, CNC, and hand tapping in most materials. High-speed, ground-thread taps are superior to cut-thread taps since the thread accuracy is held to closer tolerances.

Taper Tap:
the easiest tap to start in a hole. Requires less torque because it has more working teeth. Number of threads chamfered 7 to 10.

Plug Tap:
the most popular style for through and blind holes. Number of threads chamfered 3 to 5.

Bottoming Tap:
for tapping close to the bottom of blind holes. Number of threads chamfered 1 to 2.

Tapping & Threading

Inch Cutting Tap: Tap Drill Size

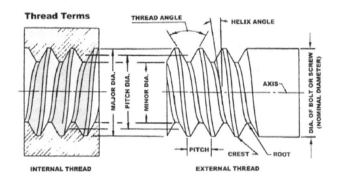

Major Dia Machine Screw Size = (Machine Screw Number x .013) + .060

Minor Dia = Major Dia – (2 x Single Thread Height)

Single Thread Height = .61343 x Pitch

Pitch = 1 / Threads Per Inch

Tap Drill Size = Major Diameter – Pitch

Tap Drill Size = Thread Dia - (0.01299 x % of Full Thread / TPI)

% of Full Thread = TPI x ((Major Dia Thread – Hole Dia) / 0.01299)

IPM (mill tapping feed rate) = RPM / TPI

IPR (lathe threading) = 1 / TPI

Thread Truncation: (Also Known as Flat on Top Of Tool)
(Rule of Thumb Formula) Thread Truncation = .125 x Pitch

Thread Gage Turns: Example
5/8 – 16 UNC To 1.75 Depth = 16 x 1.75 = 28

Tapping & Threading

Inch Roll-Form Tap (65%): Tap Drill Size

Roll-Form Tap Drill = Basic Tap Dia − ((0.0068 x % of Full Thread) / TPI)

Roll-Form Tap Drill = Basic Tap Dia − ((0.0068 x % of Full Thread) x Pitch)

Roll-Form Tap Drill = Basic Tap Dia − (0.442 / TPI)

Metric Cutting Tap: Tap Drill Size

Tap Drill (mm) = Thread Dia (mm) − ((% of Full Thread x Pitch mm) / 147.06)

Tap Drill (mm) = Thread Dia (mm) − ((% of Full Thread x Pitch mm) / 76.980)

% of Full Thread (mm) = (147.06 / Pitch in mm) x (Thread Dia − Drill Hole Dia)

IPM (mill tapping feed rate) (Metric) = RPM x Metric Pitch

IPR (lathe threading) = 1 / TPI

SMPM = RPM x Metric Pitch

Metric Roll-Form Tap (65%): Tap Drill Size

Roll Form Tap Drill Size (mm) = Basic Tap Dia − (.75 x Pitch mm x .65)

Tapping & Threading
Tread Measurement: Three-Wire Method

Best Wire Size
Dia = 0.57735 / TPI
Dia = 0.57735 x Pitch

Example: the formula above is used to find the best wire size for measuring a 1-1/4" – 12 UNC, Screw Thread.
W = 0.57735 / 12 = 0481
W = 0.57735 x .0833 = 0481

To calculate what the reading of the micrometer should be if a thread is the correct finished size, use the following constant when measuring

Unified Coarse Threads = 1.5155
American National Threads = 1.5155
Unified Fine Threads = 1.732
M = D + 3W – (1.5155 / TPI)

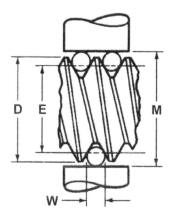

M = Mic. reading over wires
D = Diameter of thread
TPI = No. of threads per inch
W = Diameter of wire used

Example:
to find M for a 1-1/4" – 12 UNC thread, proceed as follows.
If W = .048, D = 1.250, TPI = 12
Then M =
1.250 + (3 x .048) – (1.5155 / 12) = 1.268
Then your micrometer measurement should be 1.268

Measurement over Wires:
M = (Pitch Dia + (3 x Wire Dia))
    – (.86603 x Pitch)

Note: If the best wire sizes are not available, smaller or larger wires may be used within limits. They should not be too small or too large for the thread.

Example: The best wire size for 1-1/4" – 12 UNC is .048" diameter but only a 3/64" diameter drill rod is available, which has a diameter of .0469 inch.

1-1/4" – 12 UNC
Pitch Diameter = 1.1959
(machinery handbook)
1-1/4" – 12 UNC Thread Pitch =
1 / 12 = P

M = 1.1959 – (0.86603 x P) +
(3 x wire size used)
M = 1.1959 – (0.86603 x .0833) +
(3 x .0469)
M = 1.1959 – 0.0722 + 0.1407
M = 1.2644
(this measurement will be slightly different from best wire size)

The three-wire method of thread measurement is also used for other thread forms such as Acme and Buttress. Information and tables may be found in Machinery Handbook.

Tap Drill Chart, Inch

Standard Tap Drill		Thread Size	Roll Form Tap Drill	
.0469	3/64	0-80	54	.055
.0595	53	1-64	1.65mm	.0649
.0595	53	1-72	1.7mm	.0669
.070	50	2-56	5/64	.0781
.070	50	2-64	2mm	.0787
.0785	47	3-48	43	.089
.082	45	3-56	2.3mm	.0906
.089	43	4-40	39	.0995
.0935	42	4-48	2.6mm	.1024
.1015	38	5-40	33	.113
.104	37	5-44	2.9mm	.1142
.1065	36	6-32	1/8	.125
.113	33	6-40	3.2mm	.1260
.136	29	8-32	25	.1495
.136	29	8-36	24	.152
.1495	25	10-24	11/64	.1719
.159	21	10-32	16	.177
.173	17	12-24	5mm	.1969
.180	15	12.28	8	.199
.201	7	1/4-20	1	.228
.213	3	1/4-28	A	.234
.257	F	5/16-18	7.3mm	.2874
.272	I	5/16-24	M	.295
.3125	5/16	3/8-16	S	.348
.332	Q	3/8-24	T	.358
.358	T	7/16-14	13/32	.4062
.3906	25/64	7/16-20	10.5mm	.4134
.4219	27/64	1/2-13	15/32	.4688
.4531	29/64	1/2-20	31/64	.4844
.4844	31/64	9/16-12	17/32	.5312
.5156	33/64	9/16-18	13.5mm	.5315
.5312	17/32	5/8-11	15mm	.5906
.5625	9/16	5/8-18	19/32	.5938
.6406	41/64	3/4-10	45/64	.7031
.6875	11/16	3/4-16	23/32	.7188
.7656	49/64	7/8-9	53/64	.8281
.8125	13/16	7/8-14	27/32	.8438
.8750	7/8	1-8	15/16	.9375
.9062	29/32	1-12	31/32	.9688

Tap Dill Chart, Metric

Metric Tap Drill		Metric Thread Size	Roll form Tap Drill	
Inch	mm		mm	Inch
.0295	0.75	M1 x 0.25		
.0335	0.85	M1.1 x 0.25		
.0374	0.95	M1.2 x 0.25		
.0433	1.10	M1.4 x 0.3		
.0492	1.25	M1.6 x 0.35	1.45	.0561
.0571	1.45	M1.8 x 0.35		
.0630	1.60	M2.0 x 0.40	1.85	.0709
.0689	1.75	M2.2 x 0.45		
.0807	2.05	M2.5 x 0.45	2.30	.0896
.0984	2.50	M3.0 x 0.50		.1087
.1142	2.90	M3.5 x 0.60		.1260
.1299	3.30	M4.0 x 0.70	3.7	.1437
.1457	3.70	M4.5 x 0.75		
.1654	4.20	M5.0 x 0.80		.1811
.1969	5.00	M6.0 x 1.0	5.5	.2165
.2362	6.00	M7.0 x 1.0		
.2657	6.75	M8.0 x 1.25	7.4	.2904
.2756	7.00	M8 x 1.0		.2927
.3346	8.50	M10 x 1.50	9.3	.3642
.3445	8.75	M10 x 1.25		.3658
.4016	10.20	M12 x 1.75		.4380
.4252	10.80	M12 x 1.25		.4445
.4724	12.00	M14 x 2.0	13	.5118
.4921	12.50	M14 x 1.5		.5177
.5512	14.00	M16 x 2.0	15	.5906
.5709	14.50	M16 x 1.5		.5964
.6102	15.50	M18 x 2.5		
.6496	16.50	M18 x 1.5		
.6890	17.50	M20 x 2.5	18.75	.7382
.7283	18.50	M20 x 1.5		.7539
.7677	19.50	M22 x 2.5		
.8071	20.50	M22 x 1.5		
.8268	21.00	M24 x 3.0		.8858
.8661	22.00	M24 x 2.0		
.9449	24.00	M27 x 3.0		
.9843	25.00	M27 x 2.0		
1.0433	26.50	M30 x 3.5		
1.1024	28.00	M30 x 2.0		

Metric Thread Pitch Conversion

Metric Thread Pitch	Thread Pitch In Inches	Threads Per Inch	Basic Thread Height In Inches
0.25	.00984	101.6002	.00639
0.30	.01181	84.6668	.00767
0.35	.01378	72.5716	.00895
0.40	.01575	63.5001	.01023
0.45	.01772	56.4446	.01151
0.50	.01969	50.8001	.01279
0.60	.02362	42.3334	.01534
0.70	.02756	36.2858	.01790
0.75	.02953	33.8667	.01918
0.80	.03150	31.7501	.02046
0.90	.03543	28.2228	.02301
1.00	.03937	25.4000	.02557
1.25	.04921	20.3200	.03196
1.50	.05906	16.9334	.03836
1.75	.06890	14.5143	.04475
2.00	.07874	12.7000	.05114
2.50	.09843	10.1600	.06393
3.00	.11811	8.4667	.07671
3.50	.13780	7.2572	.08950
4.00	.15748	6.3500	.10229
4.50	.17717	5.6445	.11508
5.00	.19685	5.0800	.12785
6.00	.23622	4.2333	.15344

Unified Coarse Helicoil Tap & Drill Chart

Tap Drill For Aluminum		Thread Size	Tap Drill For Magnesium Steel & Plastic	
3/32	(.0938)	2-56	(.0960)	#41
#36	(.1065)	3-48	(.1094)	7/64
#31	(.1200)	4-40	(.1200)	#31
3.4mm	(.1339)	5-40	(.1360)	#29
#26	(.1470)	6-32	(.1495)	#25
#17	(.1730)	8-32	(.1770)	#16
13/64	(.2031)	10-24	(.2055)	#5
#1	(.2280)	12-24	(.2280)	#1
H	(.2660)	1/4-20	(.2660)	H
Q	(.3320)	5/16-18	(.3320)	Q
X	(.3970)	3/8-16	(.3970)	X
29/64	(.4531)	7/16-14	(.4531)	29/64
33/64	(.5156)	1/2-13	(.5312)	17/32
37/64	(.5781)	9/16-12	(.5938)	19/32
21/32	(.6562)	5/8-11	(.6562)	21/32
25/32	(.7812)	3/4-10	(.7812)	25/32
29/32	(.9062)	7/8-9	(.9062)	29/32
1-1/32	(1.0312)	1-8	(1.0312)	1-1/32
1-11/64	(1.1719)	1-1/8-7	(1.1719)	1-11/64
1-19/64	(1.2969)	1-1/4-7	(1.2969)	1-19/64

**Max Jeffery Gilleland
Sr. Computer Tech
Assistant Instructor
Tutor
For
AutoCad
Inventor
ProEngineer
SolidWorks
Unigraphics**

Unified Fine Helicoil Tap & Drill Chart

Tap Drill For Aluminum		Thread Size	Tap Drill For Magnesium Steel & Plastic	
#37	(.1040)	3-56	(.1065)	#36
3mm	(.1181)	4-48	(.1200)	#31
#26	(.1470)	6-40	(.1495)	#25
#17	(.1730)	8-36	(.1770)	#16
#7	(.2010)	10-32	(.2031)	13/64
G	(.2610)	1/4-28	(2638)	6.7mm
21/64	(.3281)	5/16-24	(.3281)	21/64
25/64	(.3906)	3/8-24	(.3906)	25/64
29/64	(.4531)	7/16-20	(.4531)	29/64
33/64	(.5156)	1/2-20	(.5156)	33/64
37/64	(.5781)	9/16-18	(.5781)	37/64
41/64	(.6406)	5/8-18	(.6406)	41/64
49/64	(.7656)	3/4-16	(.7656)	49/64
57/64	(.8906)	7/8-14	(.8906)	57.64
1-1/64	(1.0156)	1-12	(1.0312)	1-1/32
1-1/64	(1.0156)	1-14	(1.0312)	1-1/32
1-9/64	(1.1406)	1-1/8-12	(1.1562)	1-5/32
1-17/64	(1.2656)	1-1/4-12	(1.2812)	1-9/32

Max, Assisting Student

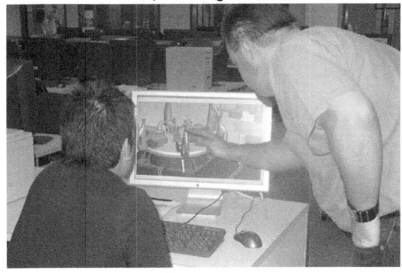

Metric Coarse Helicoil Tap & Drill Chart

Tap Drill For Aluminum		Thread Size	Tap Drill For Magnesium Steel & Plastic	
#42	(.0935)	M2.2 x 0.45	(.0935)	#42
#37	(.1040)	M2.5 x 0.45	(.1040)	#37
1/8	(.125)	M3 x 0.5	(.1250)	1/8
#27	(.1440)	M3.5 x 0.6	(.1470)	#26
#19	(.1660)	M4 x 0.7	(.1660)	#19
#5	(.2055)	M5 x 0.8	(.2090)	#4
D	(.2460)	M6 x 1	(.2500)	1/4
L	(.2900)	M7 x 1	(.2900)	L
21/64	(.3281)	M8 x 1.25	(.3320)	Q
Z	(.4130)	M10 x 1.5	(.4130)	Z
31/64	(.4844)	M12 x 1.75	(.5000)	1/2
37/64	(.5781)	M14 x 2	(.5781)	37/64
21/32	(.6562)	M16 x 2	(.6562)	21/32
47/64	(.7344)	M18 x 2.5	(.7344)	47/64
13/16	(.8125)	M20 x 2.5	(.8125)	13/16
57/64	(.8906)	M22 x 2.5	(.8906)	57/64
63/64	(.9844)	M24 x 3	(.9844)	63/64
1-3/32	(1.0938)	M27 x 3	(1.0938)	1-3/32
1-7/32	(1.2188)	M30 x 3.5	(1.2188)	1-7/32

Max, Addressing CAD Seminar

Metric Fine Helicoil Tap & Drill Chart

Tap Drill For Aluminum		Thread Size	Tap Drill For Magnesium Steel & Plastic	
21/64	(.3281)	M8 x 1	(.3281)	21/64
Y	(.4040)	M10 x 1	(.4062)	13/32
Y	(.4040)	M10 x 1.25	(.4062)	13/32
31/64	(.4844)	M12 x 1.25	(.4844)	31/64
31/64	(.4844)	M12 x 1.5	(.5000)	1/2
9/16	(.5625)	M14 x 1.5	(.5781)	37/64
41/64	(.6406)	M16 x 1.5	(.6562)	21/32
23/32	(.7188)	M18 x 1.5	(.7344)	47/64
47/64	(.7344)	M18 x 2	(.7344)	47/64
51/64	(.7969)	M20 x 1.5	(.8125)	13/16
13/16	(.8125)	M20 x 2	(.8125)	13/16
7/8	(.8750)	M22 x 1.5	(.8906)	57/64
57/64	(.8906)	M22 x 2	(.8906)	57/64
31/32	(.9688)	M24 x 2	(.9688)	31/32
1-5/64	(1.0781)	M27 x 2	(1.0938)	1-3/32
1-13/64	(1.2031)	M30 x 2	(1.2031)	1-13/64

CAD Seminar

Pipe Tap & Drill Chart

NPT Taper Pipe Tap Drill	Pipe Thread Size	NPS Straight Pipe Tap Drill
D	1/16-27	1/4
R	1/8-27	11/32
7/16	1/4-18	7/16
37/64	3/8-18	36/64
45/64	1/2-14	23/32
59/64	3/4-14	59/64
1-5/32	1 - 11 1/2	1-5/32
1-1/2	1 1/4 – 11 1/2	1-1/2
1-47/64	1 1/2 – 11 1/2	1-3/4
2-7/32	2 – 11 1/2	2-7/32
2-5/8	2 1/2 – 8	2-21/32
3-1/4	3 – 8	3-9/32

Pipe Sizes (Reference)

Nom Pipe Size	Outside Dia Inches	Sched 40 Wall Thkns	Sched 80 Wall Thkns	Sched 160 Wall Thkns
1/16	0.312	--------	--------	--------
1/8	0.405	0.068	0.095	--------
1/4	0.540	0.088	0.119	--------
3/8	0.675	0.091	0.126	--------
1/2	0.840	0.109	0.147	0.188
3/4	1.050	0.113	0.154	0.219
1	1.315	0.133	0.179	0.250
1-1/4	1.660	0.140	0.191	0.250
1-1/2	1.900	0.145	0.200	0.281
2	2.375	0.154	0.218	0.343
2-1/2	2.875	0.203	0.276	0.375
3	3.500	0.216	0.300	0.438

Feeds & Speeds

Feeds & Speeds, Formulas

Drilling:
Speeds (RPM) for drilling are calculated using the formula.

RPM = (3.82 x SFPM) / Diameter of cutter

SFPM = (RPM x Cutter Diameter) / 3.82

Turning:
Speeds (RPM) for turning are calculated using the formula.

RPM = .80 ((3.82 x SFPM) / Diameter of workpice)

SFPM = .80 ((RPM x Diameter of workpice) / 3.82)

End Mill:
Cutting Speed SFPM = 80% of drill speed.

Reaming:
Cutting Speed SFPM = 33 - 50% of drill speed.

Counter Bore:
Cutting Speed SFPM = 33% of drill speed.

Counter Sink: Cutting Speed SFPM = 33% of drill speed.

Radius Mill: Cutting Speed SFPM = 33% of drill speed

Reception After Seminar

Feeds & Speeds, Formulas

Description	Formula
Advance Per Revolution	ADVR = F / RPM
Chip Load per Tooth (Mill)	CL = IPM ÷ (N x RPM)
Cutting Time in Min. (Mill)	CTM = L / FPM
Cutting Time Seconds (Lathe)	CTS = (Dis x 60) / (FPR x RPM)
Feed Rate Per Minute (Mill)	FPM = FPT x N x RPM
Feed Rate Per Min. (Lathe)	FPM = FPR x RPM
Feed Rate Per Revolutions	FPR = IPM / RPM
Feed Per Tooth (Mill)	FPT = IPM / (RPM x N)
CL to Inches Per Minute	IPM = CL x N x RPM
Feed Rate Inch Per Minute	IPM = FPT x N x RPM
IPR to Inches Per Minute	IPM = IPR x RPM
CL to Inches Per Rev	IPR = CL x N
IPM to Inches Per Rev	IPR = IPM ÷ RPM
Distance in Minutes	L = IPM x TCm
MMPR	MMPR = IPR x 25.40
Metal Removal Rate	MMR = W x d x F
Revolutions Per Minute	RPM = (3.82 x SFM) ÷ DIA
Tapping Feed Rate	TFR = RPM ÷ TPI
Surface Feet Per Minute	SFM = (RPM x DIA) ÷ 3.82
SMPM (Metric)	SMPM = SFM x 0.3048
Mill Cutting Time In Minutes	TCm = L / IPM
Mill Cutting Time In Seconds	TCs = L / (IPM x 60)
Lathe Cutting Time In Seconds	TCs = L / (IPR x RPM x 60)

Feeds & Speeds
Abbreviations

ADV/R	=	Advance Per Revolution
°C	=	Degrees Celsius
CL	=	Chip Load Per Tooth
CTM	=	Cutting Time in Minutes (Mill)
CTS	=	Cutting Time in Seconds (Lathe)
D	=	Diameter
Dia	=	Diameter
d	=	Depth of Cut
Dis	=	Distance to Go (Inches)
F	=	Feed (Inch or Metric)
°F	=	Degrees Fahrenheit
FPM	=	Feed Per Minute (Table Travel Feedrate)
FPR	=	Feet Per Revolution
FPT	=	Feed Per Tooth
IPM	=	Inches Per Minute
IPR	=	Inches Per Revolution
L	=	Length of Cut (Inches)
MRR	=	Metal Removal Rate (cubic inch / minute)
N	=	Number of Teeth in a Cutter
RPM	=	Revolutions Per Minute
SFM	=	Surface Feed Per Minute (Metric)
SMPM	=	Surface Meters Per Minute (Metric)
MMPR	=	Millimeters Per Revolution
TCm	=	Time Cutting in Minutes
TCs	=	Time Cutting in Seconds
TPI	=	Threads Per Inch
W	=	Width of Cut

Feeds & Speeds

Chip Load (CL) Per Tooth for Mill Cutters

Cutter Type	HSS	Carbide
End Mill	.001 to .005	
Form	.001 to .005	
Saws	.001 to .002	
Plain	.002 to .004	
Stagger Tooth	.003 to .006	
Slab	.005 to .008	
Face	.010 to .012	

Increase chip loads by 50% for Aluminum Brass and Cast Iron

Drilling Speeds for Common Materials (SFPM)

Material	Hardness (Bernell)	HSS Cutter	Carbide Cutter
Aluminum	60 - 00	300 - 800	1000 2000
Brass	120 - 220	200 - 400	500 - 800
Bronze	220	65 - 130	200 – 400
Carb Steel	* 200–250		
Cast Iron **	110	50 - 80	250 –350
Copper			
Magnesium			
Molybdenum			
SST (303)			
SST (316)			
SST (410)	* 120–200		
SST (440C)	*		
Steel L/C	* 220	60 - 100	300 – 600
Steel M/C	* 229		
Steel H/C	* 240		
Titanium			

* Normalized, ** Gray

Feeds & Speeds

Cast Iron – Annealed – Scale Removed

Cast Iron Class*

Cutter	20 to 30	35 to 50	60 to 80
	Surface Feet Per Minute		
HSS	80-120	60-80	40-70
HSS Cobalt	100-150	80-120	50-100
C2 Carbide	250-500	200-450	150-350
Coated Carbide	300-700	250-550	200-450
Ceramic (no coolant)	800-1500	600-1200	500-1000

Note: Reduce de-scaling cuts by 50%. Avoid "as cast" milling operations.

* "Cast Iron Class" refers to the material PSI, i.e. 30 is a minimum of 30,000 PSI. and 80 (80,000 PSI) are nodular (ductile) cast irons which are distinguished by spheroidal graphite and higher ductility. Cutting fluids are required for these SFM speeds.

Brass & Bronze

Cutter	Anl Cast Brass	Cold Drawn Bar	Free Cutting Bar	Manganese Bronze
	Surface Feet Per Minute			
HSS	150-250	250-450	300-500	70-100
HSS Cobalt	200-350	300-600	400-700	100-150
Carbide C2	500-900	400-800	100-150	250-400
Coated Carbide	750-1500	500-1000	800-1200	350-500

Aluminum Alloys

Cutter	As Cast	Cast & Heat Treated	2024-T6 6061-T6	7075-T6
	Surface Feet Per Minute			
HSS	600 - 1000	500 - 800	500 - 800	400 - 700
HSS Cobalt	800 - 1200	600 - 1000	500 - 1200	500 - 900
Carbide C2	1200 - 2000	1000 - 1800	1000 - 1800	800 - 1500
Diamond	2500 - 8000	2500 - 6000	2500 - 6000	2500 - 4000

Diamond tools may chip or fracture if coolant is applied unevenly, so they are generally run at the lower end of the speed range without coolant.

Feeds & Speeds
Brinell Hardness (Approximate)

Steel Alloys	BHN	Stainless Steel	BHN
1018 - 1020	120-140	303	155-165
1045 - 1050	180-225	302 – 304	160-175
C12L14	155-160	316	160-175
C1215	165-170	321	145-155
StressProof	250-260	347	155-175
6150	190-205	410	150-165
8620	HR 170, CR 220	416	150-260
52100	220-225	440C	150-165
4130	255-265	15-5	310-350
4140	HR 190, CR 240 HT 290	17-4	320-340
4340	HR 230, CR 250 HT 340	17-7	360-380

Feeds & Speeds

BHN	150 – 200	205 – 230	231 – 258	264 – 301	311 – 400
R "C"	13	14 - 21	21 - 26	27-32	33-43
Cutter		Surface Feet Per Minute			
HSS	100 - 125	70 - 120	50 - 100	25 - 50	10 - 15
HSS Cobalt	125 - 150	100 - 135	70 - 125	50 - 100	35 - 80
Carbide C2	250 - 450	225 - 425	200 - 400	150 - 350	125 - 250
Coated Carbide	400 - 600	300 - 400	250 - 500	200 - 400	150 - 300
Ceramic No Coolant	900 - 1200	800 - 1000	700 - 1200	600 - 900	400 - 700

Feeds & Speeds
Machinability Comparison

Carbon Steels

1015	72%	1137	72%
1018	78%	1141	70%
1020	72%	1141 Anl	81%
1022	78%	1144	76%
1030	70%	1144 Anl	85%
1040	64%	1144 Sp	83%
1042	64%	**1212**	**100%**
1050	54%	1213	136%
1095	42%	12L14	170%
1117	91%	1215	136%

Alloy Steels

2355 Anl	70%	4620	66%
4130 Anl	72%	4820 Anl	49%
4140 Anl	66%	52100 Anl	40%
4142 Anl	66%	6150 Anl	60%
41L42 Anl	77%	8620	66%
4150 Anl	60%	86L20	77%
4340 Anl	57%	9310 Anl	51%

Stainless Steels & Super Alloys

302 Anl	45%	420 Anl	45%
303 Anl	78%	430 Anl	54%
304 Anl	45%	431 Anl	45%
316 Anl	45%	440A	45%
321 Anl	36%	15-5PH con A	48%
347 Anl	36%	17-4PH con A	48%
410 Anl	54%	A286 Aged	33%
416 Anl	110%	Hastelloy X	19%

Note: Relative machining speed based on 1212 carbon steel as 100%.
All figures are based on cold-drawn bars in as-drawn condition, except where noted.

Anl = Annealed
Sp = Stressproof
Con A = Condition A

Machinability Comparison

Tool Steels

A-2	42%	M-2	39%
A-6	33%	M-6	
D-2	27%	O-1	42%
D-3	27%	O-2	42%

Gray Cast Iron

ASTM CL20 Anl	73%	ASTM CL40	48%
ASTM CL25	55%	ASTM CL45	36%
ASTM CL30	48%	ASTM CL50	36%
ASTM CL35	48%		

Nodular Iron

60-40-18 Anl	61	80-55-06	39%
65-45-12 Anl	61		

Aluminum & Magnesium

Aluminum cold drawn	360%	Magnesium cold drawn	480%
Aluminum cast	450%	Magnesium cast	480%
Aluminum die cast	76%		

Note: Relative machining speed based on 1212 carbon steel as 100%.
All figures are based on cold-drawn bars in as-drawn condition, except where noted.

Anl = Annealed
CL20 Anl = Class 20 Annealed

Feeds & Speeds
Surface Finishes, Roughness Average

Process	Microinches	Micrometers
Sand Casting	1000 – 500	25 – 12.5
Hot Rolling	1000 – 500	25 – 12.5
Flame Cutting	1000 – 500	25 – 12.5
Sawing	1000 – 63	25 – 1.6
Forging	500 – 125	12.5 – 3.2
Planing	500 – 63	12.5 – 1.6
Drilling	250 – 63	6.3 – 1.6
EDM	250 – 63	6.3 – 1.6
Milling	250 – 32	6.3 – 0.8
Laser Cutting	250 – 32	6.3 – 0.8
Turning & Boring	250 – 16	6.3 – 0.4
Perm. Mold Casting	125 – 63	3.2 – 1.6
Investment Casting	125 – 63	3.2 – 1.6
Extruding	125 – 32	3.2 – 0.8
Cold Rolling	125 – 32	3.2 – 0.8
Reaming	125 – 32	3.2 – 0.8
Broaching	125 – 32	3.2 – 0.8
Die Casting	63 – 32	1.6 – 0.8
Grinding	63 – 4	1.6 – 0.1
Tumbling	32 – 8	0.8 – 0.2
Honing	32 – 4	0.8 – 0.1
Electro Polishing	32 – 4	0.8 – 0.1
Roller Burnishing	16 – 8	0.4 – 0.2
Polishing	16 – 4	0.4 – 0.1
Lapping	16 – 2	0.4 – 0.5
Superfinishing	8 – 1	0.2 – 0.025

Strap Clamping Force

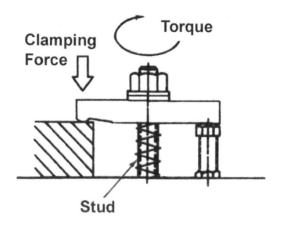

Inch Stud Size	Torque (foot – lbs)	Clamping Force (lbs)	Tensile Force In Stud (lbs)
#10-32	2	300	600
1/4-20	4	500	1000
5/16-18	9	900	1800
3/8-16	16	1300	2600
1/2-13	38	2300	4600
5/8-11	77	3700	7400
3/4-10	138	5500	11000
7/8-9	222	7600	15200
1.0-8	333	10000	20000
Metric Stud Size			
M6	4	500	1000
M8	9	900	1800
M10	20	1500	3000
M12	35	2200	4400
M16	84	4000	8000
M20	165	6300	12600
M24	283	9000	18000

Note: Clean, dry clamping stud torqued to approximately 33% of its 100,000 psi yield strength (2:1 lever ratio).

Tool Bits & Cutters

Carbon tool steel – Is used primarily to make the less expensive drills, taps, dies, and reamers. It is seldom used to make single-point cutting tools.

High-Speed Steel – The term 'high speed steel' was derived from the fact that it is capable of cutting metal at a much higher rate than carbon tool steel and continues to cut and retain its hardness even when the point of the tool is heated to a low red temperature. Tungsten is the major alloying element but it is also combined with molybdenum, vanadium and cobalt in varying amounts. Although replaced by cemented carbides for many applications it is still widely used for the manufacture of taps, dies, twist drills, reamers, saw blades and other cutting tools.

Cemented Carbide – Are also called sintered carbides or simply carbides. They are harder than high-speed steels and have excellent were resistance.

Coated Carbides – Are available only as indexable inserts because the coating would be removed by grinding. The principal coating materials are titanium carbide, titanium nitride and aluminum oxide. A very thin layer approximately .0002" of coating material is deposited over a cementen carbide insert.

Titanium Carbides – Are made entirely from titanium carbide and small amounts of nickel and molybdenum. These carbides have an excellent resistance to cratering and to heat. The high hot hardness enables them to operate at higher cutting speeds, but thy are more brittle and less resistant to mechanical and thermal shock.

Tungsten Carbide – A very hard gray compound, consisting of tungsten & carbon, made by the reaction of tungsten and carbon at high temperatures, used in making engineering dies, cutting and drilling tools, etc.

Carbide Selection – The selection of the best grade of carbide for a particular application is very important. An improper grade of carbide will result in a poor performance, It may even cause the cutting edge to fail before any significant amount of cutting has been done. Because of the many grades and the many variables that are involved in carbide cutters and there application. Recommendations should be obtained from the carbide manufacture or cutting tool producer.

Ceramic Cutting Tool Materials – Are made from finely powdered aluminum oxide particles sintered into a hard dense structure without a binder material. aluminum oxide is also combined with titanium carbide to form a composite, which is called a cermet. These materials have a very high hot hardness enabling very high cutting speeds to be used.

Cast Nonferrous Alloy – cutting tools are made from tungsten, tantalum, chromium, and cobalt plus carbon. Other alloying elements are also used to produce materials with high temperature and wear resistance. These alloys cannot be softened by heat treatment and must be cast and ground to shape. The room temperature hardness of cast nonferrous alloys is lower than for high-speed steel, but the hardness and were resistance is retained to a higher temperature.

Tool Bits & Cutters

Diamond Cutting Tools – Are available in three forms, Single-crystal natural diamonds, shaped to a cutting edge and mounted on a tool holder or on a boring bar. Polycrystalline diamond, indexable inserts made from synthetic or natural diamond powers that have been compacted and sintered into a solid mass, and chemically vapor-deposited diamond.

Single-Crystal and Polycrystalline Diamond – cutting tools are very wear-resistant, and are recommend for machining abrasive materials that cause other cutting too materials to were rapidly. Another important application of diamond cutting tools is to produce fine surface finishes on soft nonferrous metals that are difficult to finish by other methods. Surface finishes of 1 to 2 microinches can be readily obtained with single-crystal diamond tools, and finishes down to 10 microinches can be obtained with polycrystalline diamond tools.

Chemically Vapor-Deposited (CVD) Diamond – Is a new tool material offering performance characteristics well suited to highly abrasive or corrosive materials, and hard-to-machine composites. CVD diamond is available in two forms. Thick-film tools, which are fabricated by brazing CVD diamond tips to carbide substrates. And thin-film tools having a pure diamond coating over the rake and flank surfaces of a ceramic or carbide substrate.

CVD Diamond – cutting tools are recommended for the following materials; aluminum and other ductile, nonferrous alloys such as copper, brass, and bronze; and highly abrasive composite materials such as graphite, carbon-carbon, carbon-filled phenolic, fiberglass, and honeycomb materials.

Cubic Boron Nitride (CBN) – Next to diamond, CBN is the hardest known material. it will retain its hardness at a temperature of 1800 °F and higher, making it an ideal cutting tool material for machining very hard and tough materials at cutting speeds beyond those possible with other cutting tool materials.

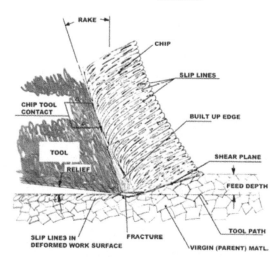

PROCESS OF CHIP FORMATION

Relief Angles HSS Tool Bits

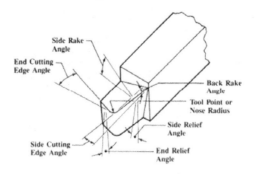

Material	Side Relief Deg	Front Relief, Deg	Back Rake Deg	Side Rake Deg
High Carbon Steel	7 – 9	6 – 8	5 – 7	8 – 10
Stainless Steel	"	"	"	"
1020, 1035, 1040	8 – 10"	8 – 10"	10 – 12"	10 – 12"
1045, 1095	7 – 9	8 – 10	10 – 12	10 – 12
1112, 1120	7 – 9	7 – 9	12 – 14	12 – 14
1314, 1315, 1385	7 – 9	7 – 9	12 – 14	14 – 16
2315, 2320, 2330, 2335, 2430, 3115, 3120, 3130	7 – 9 " "	7 – 9 " "	8 – 10 " "	10 – 12" " "
2345, 2350	7 – 9	7 – 9	6 – 8	8 – 10
3135, 3140	7 – 9	7 – 9	8 – 10	8 – 10
3250, 4140, 4340	7 – 9	7 – 9	6 – 8	8 – 10
6140, 6145	7 – 9	7 – 9	6 – 8	8 – 10
Aluminum	12 – 14	8 – 10	30 – 35	14 – 16
Bakelite	10 – 12	8 – 10	0	0
Brass free cutting	10 – 12	8 – 10	0	1 – 3
Bronze-cast	8 – 10	8 – 10	0	2 – 4
Bronze free cutting	8 – 10	8 – 10	0	2 – 4
Bronze hard phosphor	8 – 10	6 – 8	0	2 – 4
Cast Iron (gray)	8 – 10	6 – 8	3 – 5	10 – 12
Copper	12 – 14	12 – 14	14 – 16	18 – 20
Copper, hard	8 – 10	6 – 8	0	0
Copper, soft	10 – 12	8 – 10	0 – 2	0
Fiber	14 – 16	12 – 14	0 – 2	0
Formica	14 – 16	10 – 12	14 – 16	10 – 12
Nickel Iron	14 – 16	10 – 12	6 – 8	12 – 14
Micarta	14 – 16	10 – 12	14 – 16	10 – 12
Monel and Nickel	10 – 12	12 – 14	8 – 10	12 – 14
Nickel Silvers	18 – 20	10 – 12	8 – 10	0 – 2
Rubber, hard	18 – 20	14 – 16	0 - 2	0 – 2

Chip Breakers

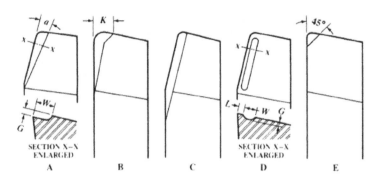

Type A
Angle a is 6 to 15-degrees, 8-degrees is a good average. Width W varies from 3/32 to 7/32 inch. Depth G may range from 1/64 to 1/16 inch. The shoulder radius equals depth G.

Type B
Is same as type A but with large nose radius and should be beveled off, as illustrated at K the width K should equal approximately 1.5 times the nose radius.

Type C
The parallel form has a tendency to force the chips against the work and damage it.

Type D
Land L for average use is approximately 1/32 inch. Width W is approximately 1/16 inch. Depth G is approximately 1/32 inch.

Type E
For light cuts and finishing cuts having a maximum depth of about 1/32 inch. This chip-breaker is a shoulder type having an angle of 45-degrees and a maximum width of about 1/16 inch.

Note: The design of chip breakers depends upon type of material as well as speeds and feeds being used. It's important in grinding all chip-breakers to give the chip-bearing surfaces a fine finish, such as would be obtained by honing. This finish greatly increases the life of the tool.

Corner Rounding Cutter

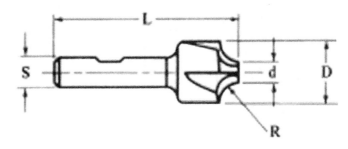

Radius R	Dia. D	Dia. d	Dia. S	Length L
1/16	7/16	1/4	3/8	2-1/2
3/32	1/2	1/4	3/8	2-1/2
1/8	5/8	1/4	1/2	3
5/32	3/4	5/16	1/2	3
3/16	7/8	5/16	1/2	3
1/4	1	3/8	1/2	3
5/16	1-1/8	3/8	1/2	3-1/4
3/8	1-1/4	3/8	1/2	3-1/2
3/16	7/8	5/16	3/4	3-1/8
1/4	1	3/8	3/4	3-1/4
5/16	1-1/8	3/8	7/8	3-1/2
3/8	1-1/4	3/8	7/8	3-3/4
7/16	1-3/8	3/8	1	4
1/2	1-1/2	3/8	1	4-1/8

Tolerances:
Radius R up to and including 1/8-inch radius, +/- 0.001
Radius R for cutters over 1/8-inch radius, +0.002 to -0.001
Diameter D = +/- .010
Diameter d =
Diameter S = -0.0001 to –0.0005
Length L = +/- 1/16

Northrop Grumman Calculator

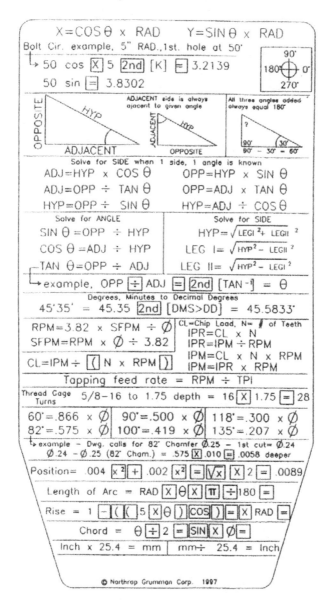

Formulas
Conversions, English & Metric
Linear Measure

1 microinch	=	0.0254 micrometers
39.37 micronich	=	1 micrometer (micron)
0.03937 inch	=	1 mm
0.3937 inch	=	10 mm or 1cm
1 inch	=	25.4 mm or 2.54 cm
3.937008 inches	=	100 mm or 10 cm
11.811024 inches	=	300 mm or 30 cm
1 foot or 12 inches	=	304.8 mm or 30.48 cm
1 foot or 12 inches	=	0.3048 meters
1yard or 36 inches	=	914.4 mm or 91.44 cm
1yard or 36 inches	=	0.9144 meters
39.37 inches	=	1 meter or 1000 mm
3.280840 feet	=	1 meter or 100 cm
0.6213712 miles	=	1 kilometer
1 mile or 5280 feet	=	1.609344 kilometers
3.106856 miles	=	5 kilometers
5 miles	=	8.04672 kilometers
6.213712 miles	=	10 kilometers
10 miles	=	16.09344 kilometers
62.13712 miles	=	100 kilometers
100 miles	=	160.9344 kilometers
621.3712 miles	=	1000 kilometers
1000 miles	=	1609.344 kilometers

Pounds – Force

1 lb-force	=	4.448222 newtons
0.2248089 lb-force	=	1 newton
1 lb-force-inch	=	0.1129848 newton-meter
8.850748 lb-force-inch	=	1 newton-meter

Formulas
Conversions, English & Metric
Square Measure

0.00155 square inch	=	1 square mm
0.1550003 square inch	=	1 square cm
1 square inch	=	645.2 square mm
1 square inch	=	6.4516 square cm
1 square foot	=	929 square cm
1 square foot	=	0.09290304 sq meters
1 square yard	=	0.836 square meters
10.76391 square feet	=	1 square meter
1.196 square yards	=	1 square meter
0.0247 acres	=	1 are = 100 sq meters
1 acre (43,560 sq feet)	=	40.46856 are
1 acre (43,560 sq feet)	=	0.4046856 hectare
2.471054 acres	=	1 hectare
0.3861 square mile	=	1 squire kilometer
1 square mile	=	2.5899 sq kilometers

Cubic Measure (Liquid)

1 U.S. quart	=	0.946 liter
1.0567 U.S quarts	=	1 liter
0.2199692 U.K. gallons	=	1 liter
0.2641720 U.S. gallons	=	1 liter
0.03531466 cubic feet	=	1 liter
1 U.S. gallon	=	3.785412 liters
1 U.K. gallon	=	4.546092 liters
1 cubic foot	=	28.31685 liters
264.2 U.S. gallons	=	1 cubic meter
1 cubic yard	=	764.559 liters

Cubic Measure (Dry)

0.06102376 cubic inch	=	1 cubic cm
1 cubic inch	=	16.38706 cubic cm
1 cubic foot	=	0.2831685 cubic meter
1 cubic yard	=	.07646 cubic meter
35.31466 cubic feet	=	1 cubic meter
1.308 cubic yards	=	1 cubic meter

Formulas
Conversions, English & Metric
Weight

1 grain	=	0.0648 gram
15.432 grains	=	1 gram
0.03527397 oz, avdp	=	1 gram
0.03215 oz, troy	=	1 gram
1 oz, avoirdupois	=	28.34952 grams
1 oz, troy	=	31.103 grams
1 pound	=	453.6 grams
1 pound	=	0.4535924 kilogram
2.204622 pounds	=	1 kilogram

Weight Per Unit Area

1 lb / sq inch	=	0.7030697 kg / sq cm
14.22334 lb / sq inch	=	1 kilogram / sq cm
1 lb / sq foot	=	4.882429 kg / sq meter
0.2048161 lb / sq foot	=	1 kg / sq meter
1 lb / sq inch	=	6.894757 kilopascals
0.1450377 lb / sq inch	=	1 kilopascals

Weight Per Unit Volume

1 lb / cubic in	=	27.67990 g / cubic cm
0.03612730 lb / cubic in	=	1 gram / cubic cm
1 lb / cubic foot	=	16.01846 kg / cubic m
0.06242797 lb / cubic ft	=	1 kg / cubic meter

Power & Heat Equivalents

778.26 foot-pounds	=	1 Btu
1 foot-bound	=	0.00128492 Btu
1 hp	=	0.7456999 kilowatts
1.341022 hp	=	1 kilowatts
1055.056 joules	=	1 Btu
1 joule	=	0.0009478170 Btu
1 foot-pound	=	1.355818 joules
0.7375621 foot-pounds	=	1 joule

Cutter Path Calculations

It is often necessary to calculate cutter paths. Even though cutter radius compensation is available on CNC machines, under certain conditions cutter comp will not give the desired results or the programmer may prefer the more direct control centerline programming affords. Centerline programming refers to the cutter path always being located on the spindle centerline (mill) or arc center of the tool nose radius (TNR) on a lathe. Formulas discussed in this section represent those most commonly used for calculating centerlines for mills and TNR centerline paths for lath applications.

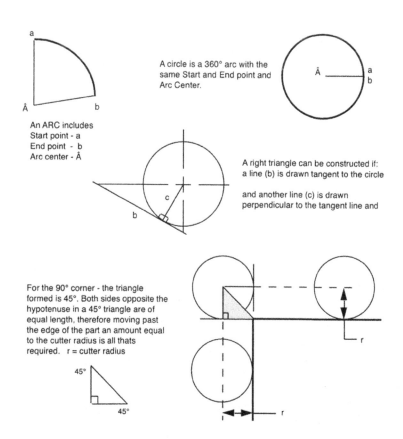

An ARC includes
Start point - a
End point - b
Arc center - Â

A circle is a 360° arc with the same Start and End point and Arc Center.

A right triangle can be constructed if:
a line (b) is drawn tangent to the circle

and another line (c) is drawn perpendicular to the tangent line and

For the 90° corner - the triangle formed is 45°. Both sides opposite the hypotenuse in a 45° triangle are of equal length, therefore moving past the edge of the part an amount equal to the cutter radius is all thats required. r = cutter radius

Cutter Path Calculations

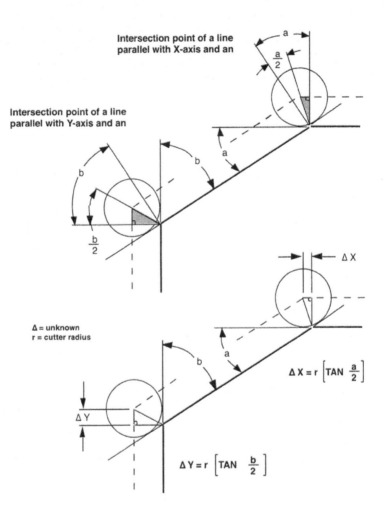

Cutter Path Calculations

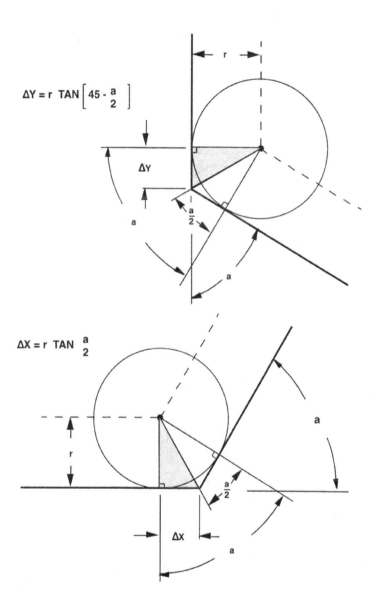

Cutter Path Calculations

Blending a Radius Into an Angle

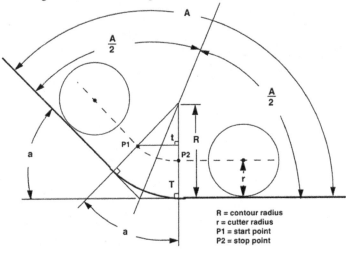

R = contour radius
r = cutter radius
P1 = start point
P2 = stop point

Angle a = 180 - Angle A   Angle A = 180 - Angle a

By constructing triangles T and t, the start and stop points (P1 & P2) can be calculated.

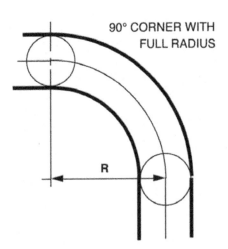

90° CORNER WITH FULL RADIUS

Outside radius: R = Part radius + cutter radius.
Inside radius: R = Part radius - cutter radius.

Tapers & Angles
Taper Per Foot & Taper Per Inch

Taper Per Foot	Included Angle			Taper Per Inch
	Deg.	Min.	Sec.	
1/16	0			
1/8	0	35	47	.010416
3/16	0	53	44	.015625
1/4	1	11	38	.020833
5/16	1	29	31	.026042
3/8	1	47	25	.031250
7/16	2	5	18	.036458
1/2	2	23	12	.041667
9/16	2	41	7	.046875
5/8	2	59	3	.052084
11/16	3	16	56	.057292
3/4	3	34	48	.062500
13/16	3	52	42	.067708
7/8	4	10	32	.072917
15/16	4	28	26	.078125
1	4	46	19	.083330
1-1/4	5	57	45	.104166
1-1/2	7	9	10	.125000
1-3/4	8	20	28	.145833
2	9	31	37	.166666
2-1/2	11	53	38	.208333
3	14	2	0	.250000
3-1/2	16	35	29	.291666
4	18	55	31	.333333
4-1/2	21	14	20	.375000
5	23	32	12	.416666
6	28	4	20	.500000

Note: Taper per inch from center line = one-half of taper per inch or Taper per inch divided by .500

Conversions, Angles
Degree, Minutes, Seconds

Convert Minutes of a Degree to a Decimal Degree:
Divide minutes by 60
Degree minutes to convert   30° 42"
Divide minutes by 60   42 / 60 = 0.7
Bring down degrees   30.7°

Convert Minutes and Seconds to Decimal Degree:
Divide seconds then minutes by 60
Degree minutes and seconds to convert   30° 41' 15"
Divide seconds by 60   15 / 60 = 0.25
Divide decimal minutes by 60   41.25 / 60 = 0.6875
Bring down degrees   30.6875°

Convert a Decimal Degree to Minutes:
Multiply decimal by 60
Decimal degree to convert   30.7°
Multiply decimal degree by 60   .07 x 60 = 42'
Bring down degrees   30° 42'

Convert Decimal Degree to Minutes and Seconds:
Multiply decimal by 60
Decimal degree to convert   30.6875°
Multiply the degree decimal by 60   0.6875 x 60 = 41.25'
Multiply decimal minutes by 60   0.25 x 60 = 15"
Bring down degrees   30° 41' 15"

Conversions, Temperature
Temperature Scales

Celsius temperature scale – registers the freezing point of water as 0 degrees C and the boiling point as 100 degrees C under normal atmospheric pressure.

Celsius (C)	1
Fahrenheit (F)	$F = C \times (9/5 + 32)$
Kelvin (K)	$K = C + 273$
Rankine (R)	$R = C \times (9/5 + 491.67)$

Fahrenheit temperature scale – registers the freezing point of water as 32 degrees F and the boiling point as 212 degrees F at one atmosphere of pressure.

Fahrenheit (F)	1
Celsius (C)	$C = (F - 32) \times 5/9$
Kelvin (K)	$C = (F - 32) \times (5/9 + 273)$
Rankine (R)	$R = F + 459.67$

Kelvin absolute temperature scale – zero occurs at absolute zero and each degree equals one Kelvin. Water freezes at 273.15 K and boils at 373.15 K.

Kelvin (K)	1
Celsius (C)	$C = K - 273.15$
Fahrenheit (F)	$F = (K - 273.15) \times (9/5 + 32)$
Rankine (R)	$R = (K - 273.15) \times (9/5 + 491.67)$
	$R = K \times (9/5)$

Rankine absolute temperature scale – using degrees the same size as those of the Fahrenheit scale, in which the freezing point of water is 491.69° and the boiling point of water is 671.69°.

Rankine (R)	1
Celsius (C)	$C = (R - 491.67) \times 5/9$
Fahrenheit (F)	$F = R - 459.67$
Kelvin (K)	$K = (R - 491.67) \times (5/9 + 273)$

Properties of Bolt Circle

The radius (r) of a bolt circle will always relate to the" hypotenuse" of a right triangle. Or the base of an oblique triangle.

Properties of Circles

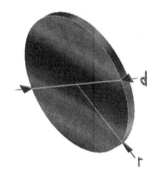

A = Area
C = Circumference
d = diameter
r = radius

$$A = \pi r^2 = 3.1416r^2 = 0.7854d^2$$

$$C = 2\pi r = 6.832r = 3.1416d$$

$$r = C + 6.2832 = \sqrt{A + 3.1416} = 0.564\sqrt{A}$$

$$d = C + 3.1416 = \sqrt{A + 0.7854} = 1.128\sqrt{A}$$

Length of arc for center angle of 1° = 0.008727d

Length of arc for center angle of n° = 0.008727nd

Properties of Circular Sector

A = area
a = angle of sector
l = arc length
r = radius

$$A = 0.5rl = 0.008727ar^2$$

$$a = \frac{57.296\ l}{r}$$

$$l = \frac{r \times a \times 3.1416}{180} = 0.01745ra = \frac{2A}{r}$$

$$r = \frac{2A}{l} = \frac{57.296\ l}{a}$$

Properties of Circular Segment

A = area
a = angle of segment
c = chord of segment
h = height of segment
l = arc length
r = radius

$$a = \frac{57.296\ l}{r}$$

$$h = r - 0.5\sqrt{4r^2 - c^2}$$

$$l = 0.01745ra$$

$$A = 0.5[(rl - c(r - h)]$$

$$c = 2\sqrt{h(2r - h)}$$

$$h = r[1.0 - \cos(a/2)]$$

$$r = \frac{c^2 + 4h^2}{8h}$$

Properties of Circular Ring

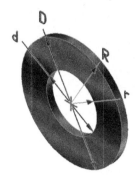

A = area
D = outside diameter
d = inside diameter
R = outside radius
r = inside radius

$$A = \pi(R^2 - r^2)$$

$$A = 2\pi(D^2 - d^2)$$

Properties of Circular Ring Sector

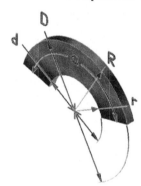

A = area
a = angle in degrees
D = outside diameter
d = inside diameter
R = outside radius
r = inside radius

$$A = \frac{a\pi}{360}(R^2 - r^2)$$

Properties of Fillet

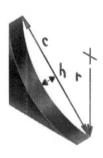

A = area
c = chord length
h = height of segment
r = inside radius

$$A = r^2 - \frac{\pi r^2}{4}$$

$$c = 2\sqrt{h(2r-h)}$$

$$h = r - 0.5\sqrt{4r^2 - c^2}$$

Trig: Right Triangle Relationships

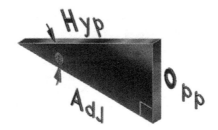

Right Triangle Relationships

$$\sin \theta = \frac{a}{c} = \frac{\text{Opp}}{\text{Hyp}} \qquad \csc \theta = \frac{c}{a} = \frac{\text{Opp}}{\text{Hyp}}$$

$$\cos \theta = \frac{b}{c} = \frac{\text{Adj}}{\text{Hyp}} \qquad \sec \theta = \frac{c}{b} = \frac{\text{Hyp}}{\text{Adj}}$$

$$\tan \theta = \frac{a}{c} = \frac{\text{Opp}}{\text{Adj}} \qquad \cot \theta = \frac{b}{a} = \frac{\text{Adj}}{\text{Opp}}$$

Complementary Relationships

$$\sin \theta = \cos (90° - \theta)$$

$$\tan \theta = \cot (90° - \theta)$$

$$\sec \theta = \csc (90° - \theta)$$

Reciprocal Relationships

$$\csc \theta = \frac{1}{\sin \theta}$$

$$\sec \theta = \frac{1}{\cos \theta}$$

$$\cot \theta = \frac{1}{\tan \theta}$$

Trig: Right Triangle

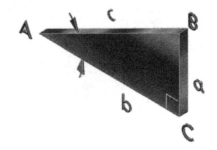

Area = (a x b) / 2

Known Sides Angles		Unknown Sides & Angles		
a - b	$c = \sqrt{a^2 + b^2}$	$A = \tan \dfrac{a}{b}$		$B = \tan \dfrac{b}{a}$
a - c	$b = \sqrt{c^2 - a^2}$	$A = \sin \dfrac{a}{c}$		$B = \cos \dfrac{a}{c}$
b - c	$a = \sqrt{c^2 - b^2}$	$A = \cos \dfrac{b}{c}$		$B = \sin \dfrac{b}{c}$
a - A	$b = \dfrac{a}{\tan A}$	$c = \dfrac{a}{\sin A}$		$B = 90° - A$
a - B	$b = a \times \tan B$	$c = \dfrac{a}{\cos B}$		$A = 90° - B$
b - A	$a = b \times \tan A$	$c = \dfrac{b}{\cos A}$		$B = 90° - A$
b - B	$a = \dfrac{b}{\tan B}$	$c = \dfrac{b}{\sin B}$		$A = 90° - B$
c - A	$a = c \times \sin A$	$b = c \times \cos A$		$B = 90° - A$
c - B	$a = c \times \cos B$	$b = c \times \sin B$		$A = 90° - B$

Trig: Oblique Triangle

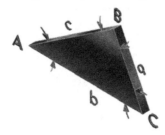

Known Sides & Angles	Unknown Sides & Angles
a-b-c	$A = \arccos \dfrac{b^2 + c^2 - a^2}{2bc}$
a-b-A	$B = \arcsin \dfrac{b \times \sin A}{a}$
a-b-C	$c = \sqrt{a^2 + b^2 - (2ab \times \cos C)}$
a-b-C	$A = \arctan \dfrac{a \times \sin C}{b - (a \times \cos C)}$
a-A-B	$b = \dfrac{a \times \sin B}{\sin A}$
a-A-C	$c = \dfrac{a \times \sin C}{\sin A}$

Area

$$A = \dfrac{a \times b \, \sin C}{2}$$

2 Angles Known

$A = 180° - (B + C)$
$B = 180° - (A + C)$
$C = 180° - (A + B)$

Trig: Oblique Triangle (cont)

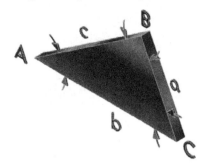

Known Sides & Angles	
	Angle B Less Than 90°
a-b-A	$B = \sin \dfrac{b \times \sin A}{a}$
a-A-C	$c = \dfrac{a \times \sin C}{\sin A}$
	Angle B Greater Than 90°
a-b-A	$B = 180° - \arcsin \dfrac{b \times \sin A}{a}$
a-A-C	$c = \dfrac{a \times \sin C}{\sin A}$

Area

$$A = \frac{a \times b \sin C}{2}$$

2 Angles Known

$A = 180° - (B + C)$
$B = 180° - (A + C)$
$C = 180° - (A + B)$

Geometric Dimension Symbols

Features	Type Of Tolerance	Characteristics	Sym.
Individual Features	"Form" Never Uses Datum	Circularity	○
		Cylindricity	⌭
		Flatness	⌗
		Straightness	—
Individual Or Related Features	"Profile" May Use Datum	Profile of Line	⌒
		Profile of Surface	⌓
Related Features	"Orientation" Always Uses Datum	Angularity	∠
		Parallelism	//
		Perpendicularity	⊥
	"Location" Always Uses Datum	Concentricity	◎
		Position	⌖
		Symmetry	⌯
	"Runout" Always Uses Datum	Runout Circular	↗
		Runout Total	↗↗

5.750	A (NEW) −A− (OLD)
Basic Dimension: A theoretically exact value used as a reference for measuring geometric characteristics and tolerances of other part features.	Datum Feature: Designates a physical feature of the part to be used as a reference to measure geometric features of other part features

Geometric Dimension Symbols
Feature Control Frame

A control frame shows a particular geometric characteristic (flatness, angularity, etc.) applied to a part feature and states the allowable tolerance. The feature's tolerance may be individual, or related to one or more datum. Any datum references and tolerance modifiers are also shown in the Feature Control Frame.

Individual Features:

▱ | .002

Related Features:

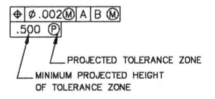

```
                    PRIMARY DATUM REFERENCE
                       SECONDARY DATUM REFERENCE
                          TERTIARY DATUM REFERENCE

        ⊕ | ⌀.002 Ⓢ | A | B | C | Ⓜ

                       MODIFIERS
                 WITHIN .002 TOTAL TOLERANCE
              TOLERANCE ZONE
           FEATURE
```

Related Features:

⊕ | ⌀.002 Ⓜ | A | B | Ⓜ
.500 Ⓟ

        PROJECTED TOLERANCE ZONE
  MINIMUM PROJECTED HEIGHT
  OF TOLERANCE ZONE

Modifiers:

Projected Tolerance Zone: An additional specification box attached underneath a Feature Control Frame. It extends the feature's tolerance zone beyond the part's surface by the stated distance, ensuring proper alignment of mating parts.

LMC = Least Material Condition:
A tolerance modifier that theoretically contains the least amount of material permitted within its

MMC = Max. Material Condition:
A tolerance modifier that theoretically contains the maximum amount of material permitted within its dimensional limits.

Geometric Dimension Symbols
Datum Target Area

Datum Targets: Callouts occasionally needed to designate specific points, lines, or areas on an actual part to be used to establish a theoretical datum feature.

DATUM TARGET AREA

 PRIMARY DATUM TARGET: 3PLACES
PRIMARY DATUM TARGET AREA
DATUM LETTER & TARGET NUMBER

 SECONDARY DATUM TARGET: 2PLACES
SECONDARY DATUM TARGET AREA
DATUM LETTER & TARGET NUMBER

 TERTIARY DATUM TARGET: 1 PLACE
TERTIARY DATUM TARGET AREA
DATUM LETTER & TARGET NUMBER

 DATUM TARGET POINT

⌀ **Cylindrical Tolerance Zone**: This symbol, commonly used to indicate a diameter, also specifies a cylindrically shaped tolerance zone in a Feature Control Frame.

Individual features *"Form" n*ever uses datum

Circularity: If a plane sliced the indicated surface perpendicular to its axis, the resulting outline must be a perfect circle, within the specified tolerance zone.

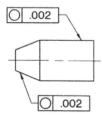

Cylindricity: All points on the indicated surface must lie in a perfect cylinder around a center axis, within the specified tolerance zone.

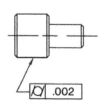

Flatness: All points on the indicated surface must lie in a single plane, within the specified tolerance zone.

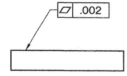

Geometric Dimension Symbols

Individual features *"Form"* never uses datum

 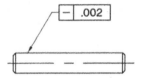

Straightness: All points on the indicated surface or axis must lie in a straight line in the direction shown, within the specified tolerance zone.

Individual or related features *"Profile"* may use datum

 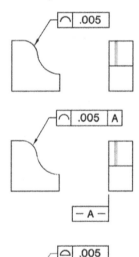

Linear Profile: All points on any full slice of the indicated surface must lie on its theoretical two-dimension profile, as defined by basic dimensions, within the specified tolerance zone. The profile may or may not be oriented with respect to a datum.

 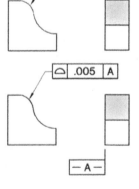

Surface Profile: All points on the indicated surface must lie on its theoretical three-dimensional profile, as defined by basic dimensions, within the specified tolerance zone. The profile may or may not be oriented with respect to a datum.

Geometric Dimension Symbols
Related features *"Orientation"* always uses datum

Angularity: All points on the indicated surface or axis must lie in a single plane at exactly the specified angle from the designated datum plane or axis, within the specified tolerance zone.

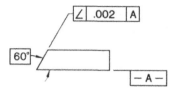

Parallelism: All points on the indicated surface or axis must lie in a single plane parallel to the designated datum plane or axis, within the specified tolerance zone.

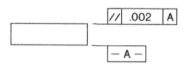

Perpendicularity: All points on the indicated surface, axis, or line must lie in a single plane exactly 90° from the designated datum plane or axis, within the specified tolerance zone.

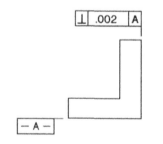

Related features *"Location"* always uses datum

Concentricity: If the surface were sliced by any plane perpendicular to the designated datum axis, ever slice's center of area must lie on the datum axis, within the specified cylindrical tolerance zone, controls rotational balance.

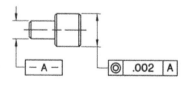

Position: The indicated feature's axis must be located within the specified tolerance zone from its true theoretical position, correctly oriented relative to the designated datum plane or axis.

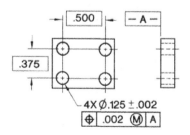

Geometric Dimension Symbols

Related features *"Location"* always uses datum

Symmetry: The indicated feature's axis must be located within the specified tolerance zone from its true theoretical position, correctly oriented relative to the designated datum plane or axis.

Related features *"Runout"* always uses datum

Runout Circular: Each circular element of the indicated surface is allowed to deviate only the specified amount from its theoretical form and orientation during 360° rotation about the designated datum axis.

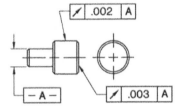

Runout Total: The entire indicated surface is allowed to deviate only the specified from its theoretical form and orientation during 360° rotation about the designated axis.

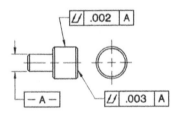

Drawing Symbols

Symbol	Description	Symbol	Description
Ø	Diameter	▼	Depth
R	Radius	□	Square Shape
CR	Controlled Radius	⟨ST⟩	Statistical Tolerance
SR	Spherical Radius	( )	Reference
SØ	Spherical Diameter	⌒	Arc Length
X	Number Of Places	▷	Conical Taper
⊔	Counterbore	▷	Slope
∨	Countersink		

Glossary

a-axis: In CNC the rotational axis about the X-axis.

Absolute Position: A CNC programming mode called by the G90 code where all tool and work-piece positions and measured from a point of absolute zero.

Acute Angle: An angle of less than 90-degrees

Angular Measure: The means by which an arc of a circle is divided and measured. This can be in degrees (360-degrees in a full circle), minutes (60 minutes in one degree), and seconds (60 seconds in one minute), or in radians. *See Radian.*

Arc: Part of the circumference of a circle or a dimension defined by an angle.

ATC: On CNC machining centers, the automatic tool changer.

Axial: Having the characteristics of an axis; situated around and in relation to an axis as in axial alignment.
In CNC related to the fundamental programmable axes of X, Y, Z, a, b, and c.

Axis: Centerline or center of rotation of an object or part; the rotational axis of machine spindle, which extends beyond the spindle and through the work-piece.

b-axis: In CNC the rotational axis around the Y-axis.

Backlash: A condition created due to clearance between a thread and nut. The amount of thread turn before a component begins to move.

c-axis: In CNC, the rotational axis around the Z-axis.

CAD: Computer-Aided-Design.

CAM: Computer-Aided-Manufacturing.

Celsius: A temperature scale used in the SI metric systems of measurement where the freezing point of water is 0-degrees and the boiling point is 100-degrees. Also called centigrade.

Chatter: Vibration of work-piece, machine, tool, or a combination of all three due to looseness or weakness in one or more of these areas.

CNC: Computer numerical control. Control of machine tools and other manufacturing equipment using computer programs.

Circumference: the periphery or outer edge of a circle. Its length is calculated by multiplying pi (3.1416) times the diameter of the circle.

Concentricity: The extent to which an object has a common center or axis. Specifically, in machine work, the extent to which two or more surfaces of a shaft rotate in relation to each other; the amount of run-out on a rotating member.

Contour: Machining an uneven but continuous path on a work-piece in two or three dimensions.

Coordinate Dimensions: A method of specifying point locations in a two-dimensional plane system defined by two perpendicular axes.

Cosine Error: A condition where the axis of a measuring instrument is out of line with the axis of the measurement to be taken, resulting in an error equal to the measuring instrument reading multiplied by the cosine of the misalignment angle.

Degrees: The circle is divided into 360-degrees, four 90-degrees quadrants. Each degree is divided into 60 minutes, and each minute into 60 seconds. Degrees are measured with protractors, optical comparators, and sine bars, to name a few methods. Degrees are also divisions of temperature.

Glossary

Diameter: Twice the radius; the length of any straight line going through the center of a figure or body, specifically a circle.

Discrimination: The degree to which a measuring instrument divides the units in which it measures. A .001-inch micrometer can discriminate to one thousandth of an inch. With a vernier, it can discriminate to one ten-thousandth of an inch (.0001 in.).

Eccentricity: A rotating member whose axis of rotation is different or offset from the primary axis of the part or mechanism. Thus, when one turned section of a shaft centers on a different axis than the shaft, it is said to be eccentric or to have "runout".

Edge Finder: A tool fastened in a machine spindle that locates the position of the work-piece edge in relation to the spindle axis.

F code word: In CNC programming, the code used to call a feed rate for milling or drilling.

G code word:
In CNC programming, a preparatory function program code that calls a particular mode of operations such as rapid traverse, linear interpolation, or circular interpolation. G codes come in groups. Each group of G codes will have a specific group number

H code word: In CNC programming, the H code refers to tool offset file numbers.

I code word: In CNC programming, and J words define arc centers.

Incremental positioning: In CNC programming a mode called by the G91 preparatory function code, where each positioning move is measured from the point where the tool is presently located.

Interpolation: In CNC programming, a "best Fit" tool path along an angular, circular, or helical programmed path.

IPM rate: In machining, a feed rate measured in inches per minute.

IPR rate: In machining, a feed rate specified in inches per revolution of the machine spindle.

J code word: In CNC programming, The and J words specify arc center locations.

Lead: The distance a thread or nut advances along a threaded rod in one revolution.

M code word: In CNC programming, M codes are miscellaneous function codes such as M06 for tool change or M03 for spindle stop.

Machinability: The relative ease of machining, which is related to the hardness of the material to be cut.

Model Commands: In CNC programming, model commands once invoked remain in force until canceled or replaced with another G code from the same group. All G codes except group 00 are model.

Non-Model Commands: In CNC programming non-model commands once invoked remain in force only in the calling block, and are immediately forgotten by the control. All G codes in group 00 are Non-model.

Nose Radius: Refers to the rounding of the point of a lathe cutting tool. A large radius produces a better finish and is stronger than a small one.

Origin: The intersection of the X, Y, and Z-axes.

Parallax Error: An error in measurement caused by reading a measuring device, such as a rule, at an improper angle.

Glossary

Radian: an angle of 57.2958° degrees, equal to the angle at the center of a circle formed by an arc equal in length to the radius of the circle.
180° / 3.1416 = 57.2958

Thread Axis: The centerline of the cylinder on which the thread is made.

Thread Chasing: in machining terminology, chasing a thread is making successive cuts in the same groove with a single-point tool. Also done when cleaning or repairing a damaged thread.

Thread Crest: Outer edge (point or flat) of a thread form.

Thread Engagement: the distance a nut or mating part is turned onto the thread.

Thread Fit: Systems of thread fits for various thread forms range from interference fits to loose fits.

Thread Lead: The distance a nut travels in one revolution. The pitch and lead are the same on single-lead threads but not on multiple-lead threads.

Thread Pitch: The distance from a point on one thread to a corresponding point on the next thread.

Thread Relief: Usually an internal groove that provides a terminating point for the threading tool.

Thread Taper: A thread made on a taper such as a tapered pipe thread.

Tool Offset: In CNC programming, the distance from the tool end to the work-piece.

X – Axis: On CNC machining centers the table axis. On CNC turning centers, the cross slide axis.

X – Y Plane: The plane formed by the X and Y-axis.

X – Z Plane: The plane formed by the X and Z-axis.

Y – Axis: On a CNC machining center, the saddle axis.

Y – Z Plane: The plane formed by the Y and Z-axis.

Z – Axis: On CNC machine tools in general, the spindle axis.

Zero-Index: Also zero point. The point at which micrometer dials on a machine are set to zero and the cutting tool is located to a given reference, such as a work-piece edge.

Haas CNC Notes

1) A Decimal Equivalent Chart Can be displayed on a Haas control by pressing the HELP/CALC button, and then selecting the Drill Table Tab. Use the jog handle or cursor keys to scroll through the chart.

2) Spindle Command: You stop or start the spindle with CW or CCW (FWD and REV on a Lathe) any time you're at Single Block stop or a Feed Hold. When you restart the program with Cycle Start, the spindle will be turned back on to the previously defined speed.

3) When in EDIT or MEM mode you can select and display another Program quickly by simply entering the program number (Onnnnn) and pressing the cursor arrow down key.

4) Metric Tap and Drill Sizes based on 77% full metric thread can be displayed on a Haas control by pressing the HELP/CALC button, and then selecting the Drill Table tab.

5) Clearing Current Commands Values on a Haas, the values in the CURNT COMDS display pages for Tool Life, Tool Load and Timer registers can be cleared by cursor-selecting the one you wish to clear and pressing ORIGIN. To clear everything in a column, cursor to the top of that column (onto the title) and press ORIGIN.

6) To Rapid an Axis Home you can rapid all axis to machine zero by pressing the HOME G28 key. You can also send just one axis (X, Y, Z, A or B) to machine zero in rapid motion. Enter the X, Y, Z, A or B, then press HOME G28 and that axis alone will rapid home. *CAUTION!* There is no warning to alert you of a possible collision.

7) In the Offset display on a Haas you can zero all offsets at once by pressing ORIGIN. It will respond with ZERO ALL (Y/N). if you press Y, it will zero all offsets. *CAUTION!* You can't undo this command.

8) When Setting 32 on a Haas machine is set to IGNORE, Then all commands for turning coolant on or off will be ignored. The coolant can still be turned on and off manually with the COOLNT button.

9) Haas training manuals:

and other information may be downloaded for free from the Haas website (www.haascnc.com/training). On the main page, click on Customer Service. A menu will appear below, Click on Haas Training. Haas training information will be listed for you to access and download. Information will be updated periodically with new training material and updates.

10) You can edit programs on a Haas while a program is running, using Background (BG) Edit. When running a program in MEM mode from the Program display, enter the program number you want to edit (0nnnnn) and press F4 to get to BG Edit. You can then simple edits with INSERT, ALTER, DELETE, and UNDO. You can also block edit in BG Edit. Put the cursor on the first line of a program block and press F1, then scroll down to the last line and press F2. The entire block will be highlighted. Cursor to the desired point in the program, and press INSERT to copy the block there, ALTER to move the block, DELETE to delete it, and undo to remove the highlighting. Press F4 again to exit BG Edit. You can also use BG Edit to create a new program or edit the program that's running. When you BG Edit the program that's running, the changes will not take effect until next cycle.

11) Setting 9 on a Haas allows you to change between inch and millimeter dimensioning.

Haas CNC Notes

12) To Zero the POS-OPER: this display is used for reference only. Each axis can be zeroed out independently, to then show the position relative to where you selected to zero that axis. To zero out a specific axis, PAGE UP or PAGE DOWN in the POST display to the POS-OPER large-digit display page. When you Handle Jog the X, Y or Z axis and then press ORIGIN, the axis that is selected will be zeroed. Or, you can press an X, Y or Z letter key then press ORIGIN to zero that axis display. You can also press the X, Y or Z key and enter (X2.125), then press ORIGIN to enter the number in that axis display.

13) When a program is running in Memory mode from the PGRM display on a Haas, you can scroll through the program that is running by pressing F4 to get a Program Review display. You can then scroll through the entire program with the cursor keys. Press F4 again to exit Program Review.

14) Setting 22 on a Haas, Can Cycle Delta Z, defines the distance above the previous peck that a tool will rapid back to during a Mill and Lathe G83 peck drill or the amount it pulls back in a G74 and G75 Lathe Grooving cycle. It also defines the distance the tool retracts to break the chip in a Mill G73 peck drill cycle.

15) Transferring Simple Calculations in the Haas Calculator display, the number in the simple calculator box (upper left corner) can be transferred to any cursor-selected data line on the page in EDIT or MDI. Cursor to the register to which you wish to transfer the calculator number, and press F3.

16) On a Haas, you can use the DIST-TO-GO screen to quickly zero out the Position display for a reference move. When in Handle Jog mode and in the Position Display, press any other mode key (EDIT, MEM, etc.) and then go back to Handle Jog. This will zero out all axis on the DIST-TO-GO display, and begin showing the distance moved.

17) On a Haas, it's easy to transfer a program from MDI and save it to your list of programs. In the MDI display, make sure that the cursor is at the beginning of the MDI program. Enter a program number (Onnnnn) that's not being used. Then press Alter and this will transfer the MDI data into your List of Programs under that program number.

18) Chip Conveyor: The chip conveyor on a Haas can be turned on or off when a program is running using the control keys or in the program using M-Codes. The M-Code equivalent to CHIP FWD is M31, and CHIP STOP is M33. You can set the Conveyor Cycle time (in minutes) with Setting 114, and the Conveyor On-Time (in minutes) with Setting 115.

19) Setting 36 PROGRAM RESTART: When it is ON, you are able to start a program from the middle of a tool sequence. You cursor onto the line you want to start on and press CYCLE START. It will cause the entire program to be scanned to ensure that the correct tools, offsets G-Codes, and axis positions are set correctly before starting and continuing at the block were the cursor is positioned. Although you can leave this setting ON all the time, it may cause the machine to perform certain activities unnecessarily, so it's best to turn it OFF when you're done using it

20) Memory Lock Keyswitch: This is a Haas machine option that prevents operators from editing or deleting programs, and from altering settings. Since the keyswitch (if installed) locks the settings, it also allows you to lock areas within settings. Setting 7 locks Parameters 57, 209, and 278 lock other features. Setting 8 locks all programs. Setting 23 locks O9xxx programs. Setting 119 locks offsets. Setting 120 locks macro variables. In order to edit or change these areas, the Keyswitch (if installed) must be unlocked and its setting turned off.

Haas CNC Notes

21) Tool Life Management: In the CURNT COMDS display on a Haas you can PAGE DOWN to the Tool Life management page. On this page the Tool Usage register is added to every line that tool is called up in the spindle. You enter the of times you want that tool to be used in the Alarm column. When the Usage number for that tool reaches the number of uses in the alarm column, it will stop the machine with an alarm. This will help you monitor tools to prevent them from breaking, and prevent parts being scrapped.

22) Setting 103: CYC START / FH SAME KEY. This is really good to use when you're carefully running through a program on a Haas. When this setting ON, the CYCLE START button functions as a Feed Hold key as well. When CYCLE START is pressed and held in, the machine will run through the program; when it's released, the machine will stop in a Feed Hold. This gives you much better control when testing a new program. When you're done using this feature, turn it off. This setting can be changed while running a program. It cannot be on when Setting 104 is on; when one of these settings is turned on, the other will automatically turn off.

23) Setting 104: JOG HANDL TO SNGL BLK. When setting 104 is on and a program is running in MEM mode in the program or Graphics display, pressing the SINGLE BLOCK key allows you to cycle through your program one line at a time, whether the machine is running or your in Graphics. First press the CYCLE START button, and then each counterclockwise click of the jog handle will step you through the program line by line. Turning the handle clockwise will cause a Feed Hold this setting can be changed while running a program. It cannot be on when Setting 103 is on; when one of these settings is turned on, the other will automatically turn off.

24) Advanced Editor Quick Cursor Arrow: You can call up a cursor arrow with which to scroll through your program quickly, line by line, when you're in the Advanced Editor. For quick cursor arrow, press F2 once; then you can use the Jog Handle to scroll line by line through the program. To get out of this Quick-Cursor mode and remain where you are in the program, just press the UNDO key.

25) Duplicating a Program in LIST PROG: in the LIST PROG mode, you can duplicate an existing program by cursor-selecting the program number you wish to duplicate, typing in a new program number (Onnnnn), and then pressing F2 (on older machines, press F1). You can also go to Advanced Editor menu to duplicate a program, using the PROGRAM menu and the DUPLICATE ACTIVE PROGRAM item.

26) Tool Load Management: Press the PAGE UP or PAGE DOWN key in CURNT COMDS to page to the Tool Load Page. Spindle load condition can be defined for a particular tool, and the machine will stop if it reaches the spindle load limit defined for that tool. A tool overload condition can result in one of four actions by the control.

The action is determined by Setting 84. ALARM will generate an alarm when overload occurs; FEED HOLD will stop with a Feed Hold when overload occurs; BEEP will sound an audible alarm when overload occurs; or AUTOFEED will automatically decrease the feed rate. This will also help you monitors tools.

27) Jog Keys: You can select an axis for jogging on a Haas by entering the axis letter on the input line and then pressing the HANDLE JOG button.

Haas CNC Notes

28) Leaving Messages: You can enter a message in the MESGS display for the next operator, or yourself. It will be the first screen when you power up the machine, if there are no alarms other than the usual 102 SERVOS OFF alarm. If the machine was powered down using EMERGENCY STOP, the MESGS display will not show up when you turn the machine on again. Instead, the control will display the active alarm generated by the emergency stop. In this case, you would have to press the ALARM/MESGS key to view a message. It is not necessary to hit EMERGENCY STOP when you power down a Haas machine.

29) Send and Receive Offsets, Settings, Parameters, Macro Variables to/ from Disk, USB, etc.: You can save offsets, settings and parameters to a floppy, hard disk, USB or via RS-232. Press LIST PROG, select DESTINATION and then select an OFSET, SETNG, PARAM or Macro Variables (PAGE DOWN from CURNT COMDS) display page. Type in a file name and then press F2 to write that display information to disk. Entering the file name and pressing F3 will read that file from the disk into the selected display.

30) Send ALL Programs, Offsets, Settings, Parameters, alarm history, and Macro Variables at once to/from Disk, USB, etc.: Under a filename and a separate file name extension: a) Select LIST PROG b) Select POSIT (you will see the F2 and F3 prompt at the bottom left of the display) c) Enter a filename without extension. d) Press the F2 key. e) All six areas will be saved under the same filename with an extension (.pgm, .ofs, .set, etc.) f) you can read these files into the control by entering the filename and pressing F3.

Exercises

CNC Mill Exercises:
Bolt Hole Location
Trig Cutter Path
Positioning: Absolute & Incremental
Linier Interpolation

CNC Mill Programming:
Plate: With Triangular Hole Pattern
Plate: With Three Grooves

CNC Mill Programming:
Plate: Machined Both Sides, Side 1
Plate: With Bolt Circles
Plate: Machined Both Sides, Side 2
Plate: Happy Face

Editors: Suggested Reading

$$RPM = (3.82 \times SFPM) / Dia$$

$$FPT = IPM / (RPM \times N)$$

$$SFPM = (RPM \times Dia) / 3.82$$

$$Pitch = \frac{1}{TPI}$$

$$IPM = IPR \times RPM$$

Exercise
Bolt Hole Location

Often bolt hole circles, or a series of holes located on an arc, must be placed in a machined part. Typically only the number of holes and the diameter of the bolt circle are provided; which requires the calculation of the X, Y coordinates for each hole. With the help of a calculator and using the laws of right triangles, the locations can be quickly and easily calculated.

As illustrated below, construct a right triangle using the bolt circle centerlines as the base and top of the triangle intersecting the center of a hole.

The hypotenuse of the triangle is equal to the bolt circle radius (R). angle (A) is determined by the number of holes in the bolt circle. Sides "a" and "b" are calculated and used to determine the X and Y coordinate for each hole.

The triangle can be moved (broken lines) to each hole, therefore, after sides "a" and "b" have been calculated the values can be used to locate the other holes on the bolt circle.

The Process to find sides "a" and "b" are simple:

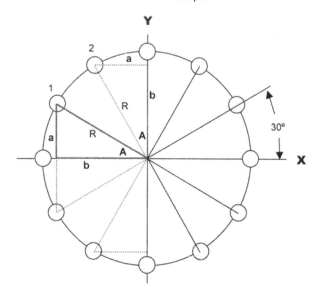

Example:

Side a = sine of angle A x radius
Side b = cosine of angle A x radius

Find the X, Y location for holes 1 and 2.

Known Values:
Bolt circle diameter = 4.000 inches
X, Y zero is the bolt circle center.
Angle "A" = 360° divided by number of bolt holes (12) = 30°

Calculations:

Side "a" = sine 30° x 2.000 = 1.0000
Side "b" = cosine 30° x 2.000 = 1.7320

Answers:

Hole 1 =   X-1.7320   Y1.0000
Hole 2 =   X-1.0000   Y1.7320

Exercise
Bolt Hole Location

Instructions: Calculate the coordinates for each hole. X, Y position. Zero is the center of the bolt-hole circle. Hole sequencing is in the clockwise direction.

Hole #1 is located at the 12 o'clock position or the closest hole located to the 12 o'clock position

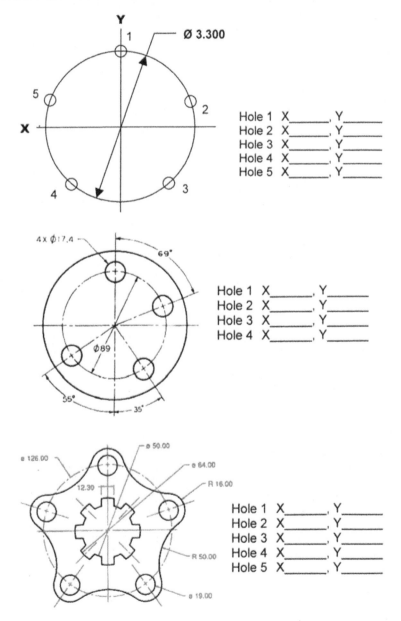

Hole 1 X_____, Y_____
Hole 2 X_____, Y_____
Hole 3 X_____, Y_____
Hole 4 X_____, Y_____
Hole 5 X_____, Y_____

Hole 1 X_____, Y_____
Hole 2 X_____, Y_____
Hole 3 X_____, Y_____
Hole 4 X_____, Y_____

Hole 1 X_____, Y_____
Hole 2 X_____, Y_____
Hole 3 X_____, Y_____
Hole 4 X_____, Y_____
Hole 5 X_____, Y_____

Exercise
Bolt Hole Location

Instructions: Calculate the coordinates for each hole. X, Y position. Zero is the center of the bolt-hole circle. Hole sequencing is in the clockwise direction.

Hole #1 is located at the 12 o'clock position or the closest hole located to the 12 o'clock position

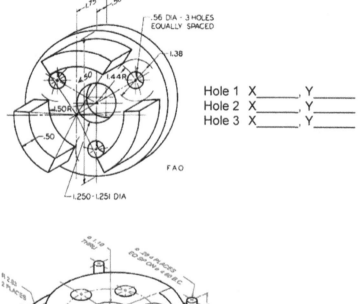

Hole 1 X_____, Y_____
Hole 2 X_____, Y_____
Hole 3 X_____, Y_____

Ø .590
X6
EQ SPACED ON
Ø 2.620

Hole 1 X_____, Y_____
Hole 2 X_____, Y_____
Hole 3 X_____, Y_____
Hole 4 X_____, Y_____
Hole 5 X_____, Y_____
Hole 6 X_____, Y_____

Exercise
Trig Cutter Path

Instructions:

Calculate the X – Y Coordinates to mill the part shown below.

Material: 6061-T6 Aluminum x 1/2" Thk.

Cutter: 0.500 Dia. End Mill, HSS

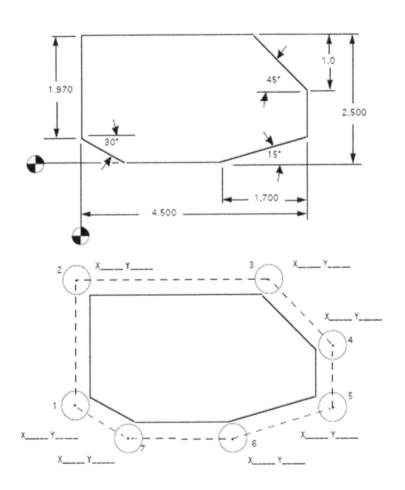

Exercise
Trig Cutter Path

Instructions:

Calculate the X – Y Coordinates to mill the part shown below.

Material: A2 Tool Steel x 3/4" Thk.

Cutter: 1.000 Dia. End Mill, Carbide

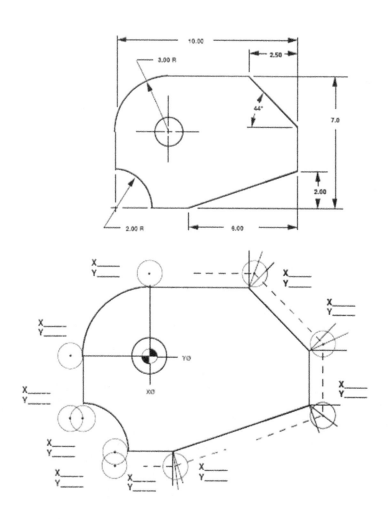

Exercise
Trig Cutter Path

Instructions:

Calculate the X – Y Coordinates to mill the part shown below.

Material: 304 Stainless Steel x 1/4" Thk.

Cutter: 0.3125 Dia. End Mill, Carbide

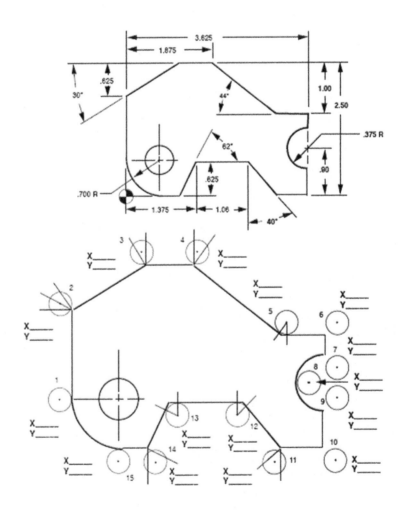

Exercise
Positioning, G90 Absolute & G91 Incremental

G90 Absolute:
What is the value in X and Y for each hole in absolute G90 positioning when each move is defined from a single fixed part zero point of an X0 Y0 origin Point

Note: Each square of the grid below represents 0.250 x 0.250 inch.

G91 Incremental:
What is the value in X and Y for each hole in incremental G91 positioning when each move is defined from the previous position and the zero point shifts with the new position.

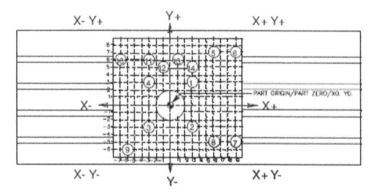

G90, Absolute Positioning

PT1 = X_____ Y_____
PT2 = X_____ Y_____
PT3 = X_____ Y_____
PT4 = X_____ Y_____
PT5 = X_____ Y_____
PT6 = X_____ Y_____
PT7 = X_____ Y_____
PT8 = X_____ Y_____

G91, Incremental Positioning

From PT8 to PT9 = X_____ Y_____
From PT9 to PT10 = X_____ Y_____
From PT10 to PT11 = X_____ Y_____
From PT11 to PT12 = X_____ Y_____
From PT12 to PT13 = X_____ Y_____
From PT13 to PT14 = X_____ Y_____

Exercise
Program Interpolation

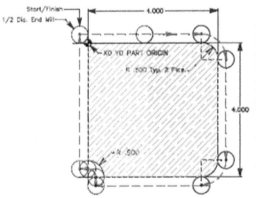

Cutter will be a .500 dia. end mill. Start contour from the top left corner of part and mill around outside of part .625 deep. When defining a circular move (G02 or G03) you can use either an I,J,K or an R command but not both. *Cutter Compensation is not being used in this exercise so Mill around outside of part with end mill defining the very center of cutter to position around part.*

"I" = X axis incremental distance and direction from the start point to the arc center.
"J" = Y axis incremental distance and direction from the start point to the arc center.

```
O00010 (INTERPOLATION EXERCISE)
T1 M06 (1/2 DIA. 4 FLT END MILL)
G90 G54 G00 X_____ Y_____ (Start/Finish)
S1520 M03
G43 H01 Z0.1 M08
G01 Z_____ F50. (Fast feed down to depth, non-cutting move)
X_____ F12. (Feed end mill over to the top right radius and continue around
part, to the end, defining center of tool. Cutter compensation is not being used here.)
G0__ X_____ Y_____ R_____ (or use I_____ and J_____ instead of R)
G0__ Y_____
G0__ X_____ Y_____ R_____ (or use I_____ and J_____ instead of R)
G0__ X_____
Y_____
G0__ X_____ Y_____ R_____ (or use I_____ and J_____ instead of R)
G0__ X_____
Y_____
G00 Z1. M09
G28 G91 Z0. M05
M30
```

Exercise
Program Plate, With Triangular Hole Pattern

Program Objective: In Completing this program you will create.

Rectangular coordinates (quadrant #1)
Rapid traverse moves.
Linear moves with feed rate.
A canned cycle for drilling (G81)
Spindle speeds.
Tool length compensation.

Miscellaneous functions such as
M3 Spindle on FWD
M5 Spindle Off
M30 End of Program

1) Fill in the following Information.
Spindle speed =
Feed rate =
Cutter clearance =

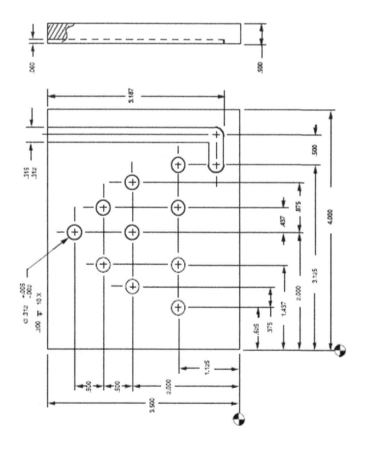

Exercise
Program Plate, With Triangular Hole Pattern

2) Calculate and list the cutter locations on the coordinate worksheet.

Other Instructions and notes:

3) Recheck your calculations.

4) Write a complete program using your calculations from your coordinate worksheet.

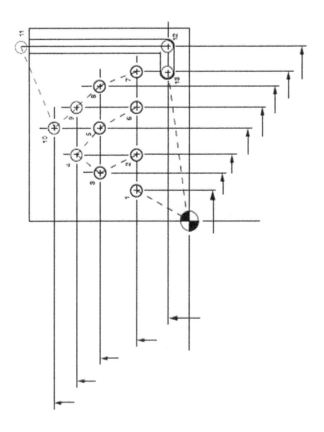

Exercise
Program Plate, With Three Grooves

Program Objective: This program will help reinforce the new procedures.

New procedures.
Linear interpolation for milling angles.
Utilizing initial and R planes with G81.
Milling open-ended pockets.

1) Fill in the following Information.
Spindle speed =
(use 300 SFPM)

Feed rate =
(use .003 chip load)

Cutter clearance =

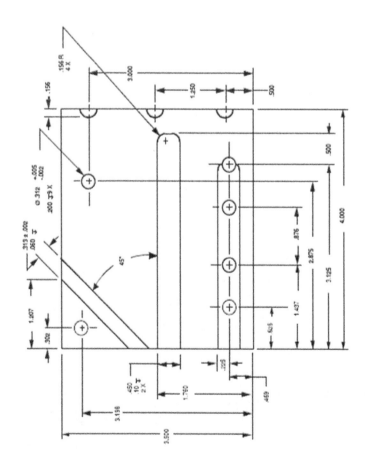

Exercise
Program Plate, With Three Grooves

2) Calculate and list the cutter locations on the coordinate worksheet.

3) Recheck your calculations.

4) Write a complete program using your calculations from your coordinate worksheet.

Other Instructions and notes:
Position #7 leave 0.050" for clean up.

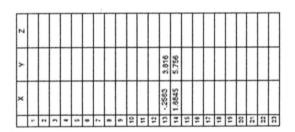

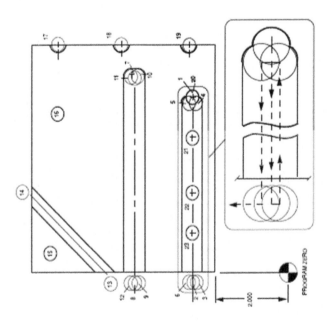

Exercise
Program Plate, Machined Both Sides, Side 1

Program Objective: This program will help reinforce the new procedures.

New procedures.
Rectangular coordinate moves in quadrant #4. Setting up tool changes.
Calculating Z depths for drills.
A canned cycle for peck drilling (G83)
Pocking milling using two methods.
Tool change position.

1) Fill in the following Information.
Spindle speed =
(use 300 SFPM)

Feed rate =
Cutter clearance = 0.100"
Peck Depth =
3 Pecks =
2 Pecks =

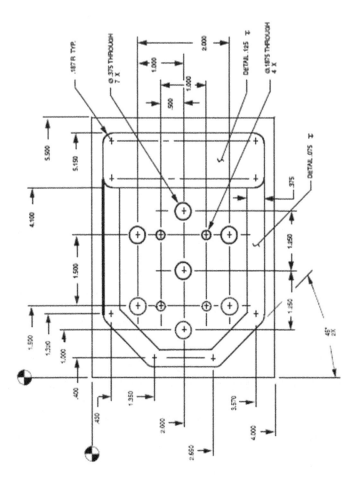

Exercise
Program Plate, Machined Both Sides, Side 1

2) Calculate and list the cutter locations on the coordinate worksheet.

3) Recheck your calculations.

4) Write a complete program using your calculations from your coordinate worksheet.

Other Instructions and notes:

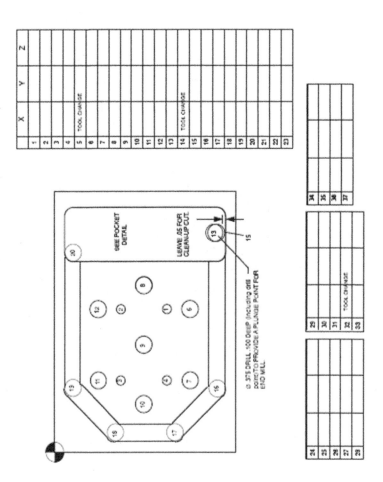

Exercise
Program Plate, Machined Both Sides, Side 1

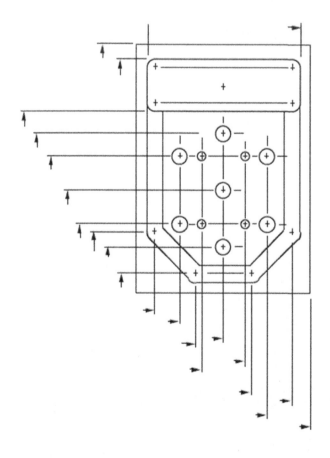

Exercise
Program Plate, Machined Both Sides, Side 1

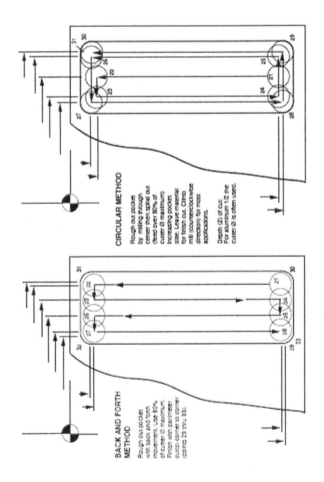

Exercise
Program Plate, With Bolt Circles

Program Objective: This program will help reinforce the new procedures.

New procedures.
Rectangular coordinate moves in all 4 quadrants.
Bolt hole circle layout.
A canned cycle for counterboring (G82)

1) Fill in the following Information for all three tools.

Tool #   Description   RPM   Feed

a)

b)

c)

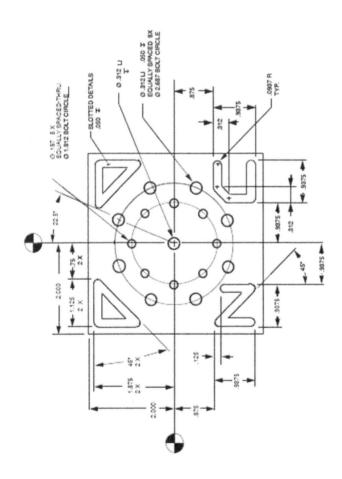

Exercise
Program Plate, With Bolt Circles

2) Calculate and list the cutter locations on the coordinate worksheet.

3) Recheck your calculations.

4) Write a complete program using your calculations from your coordinate worksheet.

Other Instructions and notes:
See Machinery's Handbook for bolt-hole circle charts.

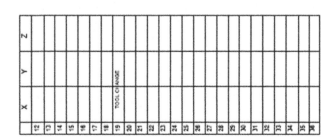

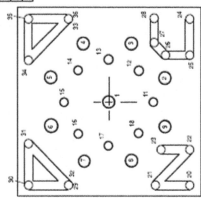

Exercise
Program Plate, Machined Both Sides, Side 2

Program Objective: This program will help reinforce the new procedures.

New procedures.
Cutter compensation.
Circular interpolation using R command.
Circular interpolation using I & J command.
Cycle repeat command.

1) Fill in the following Information for both tools.

    Tool # Description RPM Feed

a)

b)

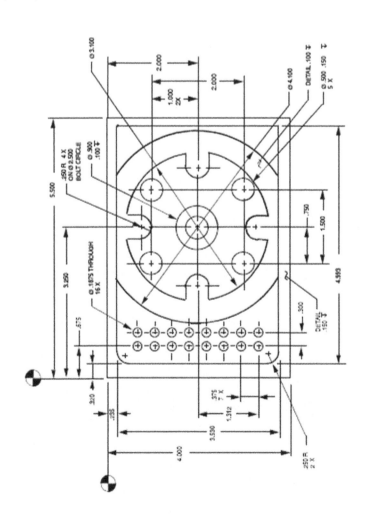

Exercise
Program Plate, Machined Both Sides, Side 2

2) For points #1 through #7, use cutter compensation (G41 & G42)

3) Use the circular radius codes as noted.

4) 0.500" Dia. counterbores use G82.

5) For 0.1875" Dia holes use G83 with L repeat counter.

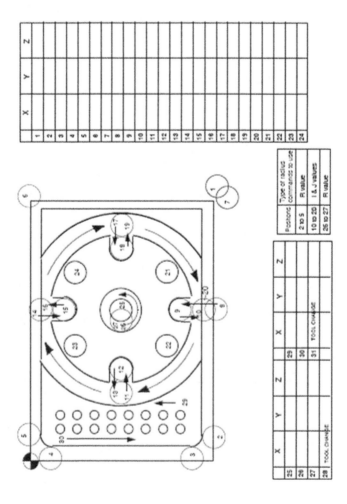

Exercise
Program Plate, Machined Both Sides, Side 1 & 2

6) Calculate and list the cutter locations on the coordinate worksheet.

7) Recheck your calculations.

8) Write a complete program using your calculations from your coordinate worksheet.

Other Instructions and notes:
See Machinery's Handbook for bolt-hole circle charts.

CNC PROJECT
Top and bottom views after machining one side with Program #3 and the opposite side with Program #5

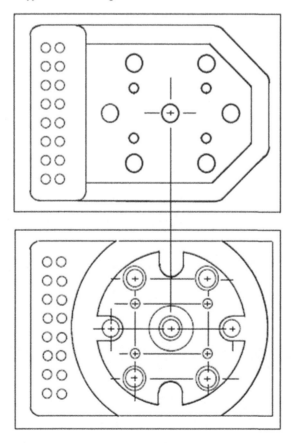

Exercise
Program Happy Face

Instructions: To help understand the use of circular interpolation commands, write a program to machine the happy face shown below. Use the tools and cutter paths illustrated on the next page.

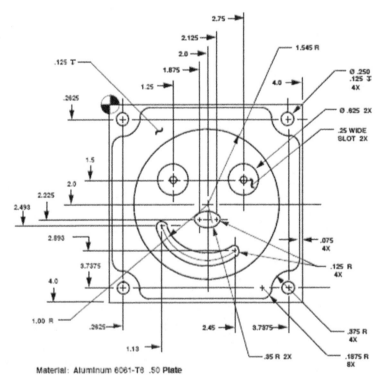

Material: Aluminum 6061-T6 .50 Plate

Note: All depths are from top surface.
Eyes, nose, & mouth all .06 deep.

Exercise
Program Happy Face

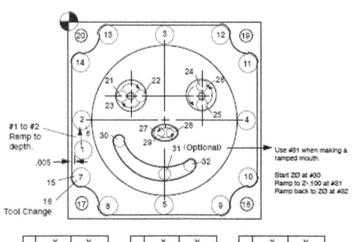

Editors
Suggested Reading

CNC Programming Handbook,
by Peter Smid
Industrial Press, Inc.
New York, NY 10018
www.industrialpress.com
(CNC simulation software with book)

CNC Programming Techniques,
by Peter Smid
Industrial Press, Inc.
New York, NY 10018
www.industrialpress.com

Jig and Fixture Handbook
Carr Lane Manufacturing Co.
4200 Carr Lane Ct., PO Box191970
St. Louis, Missouri 63119
www.carrlane.com

The CNC Workshop
by Frank Nanfara, Tony Uccello, and
Derek Murphy
SDC Publications, Mission, Kansas
www.schroff.com
(CNC simulation software with book)

The Starrett Book For Student Machinists
The L. S. Starrett Company
World's Greatest Toolmakers
Athol, Massachusetts, 01331
www.starrett.com

De Anza College
MCNC Program Courses for Certificate or AS Degree
Course Description

MCNC 64: Manufacturing Materials & Processes

Applied materials and process analysis, materials and process selection techniques. The role of metals, polymers, ceramics and composites in the casting, molding, forging, forming, machining, joining, and heat and surface treatment processes.

MCNC 71: Intro to Machining & CNC Processes

Manufacturing lab safety. Precision measuring tools and practices. Basic manual machine operations: pedestal grinders, drill presses, saws, lathes, and milling machines. Threads: types, applications and use of taps and dies. Cutter speed and feed calculations.

MCNC 72: Applied GD&T and CMM Measuring

Interpretation of specifications and inspection procedures related to current ASME Y14.5 GD&T standards. Applications and capabilities of precision measuring tools, including the computer-aided CMM, used in manufacturing environments to inspect discrete complex parts.

MCNC 75A: Introduction To CNC

Introduction to mill tool path programming using G & M code format. CNC systems and components including machine controller functions and operations. Program entry, editing, and back plotting. Calculation for mill and lathe cutter compensation.

MCNC 75B : CNC Advanced Mill

Introduction to tool path programming using word address format, including coordinate system, cutter compensation and canned cycles. Advanced mill programming; sub programs. Work coordinate system and use of macros. Program entry and editing, and back plotting, machine controller functions and operations.

MCNC 75C: CNC Advanced Lathe & 4-axis Mill

CNC lathe tool path programming using G & M code format, including tool orientation and compensation and canned cycles. Programming for CNC horizontal machining centers and 4-axis rotary tables. Horizontal machining center and lathe controller functions, setup and operations. Fixture designing for mills and lathes.

MCNC 76C: CAD/CAM Mastercam (Introduction)

Three-axis mill programming; creating part geometry, defining tools and tool paths, and using post-processors to produce word-address format programs.

MCNC 76H ; CAD/CAM Mastercam (3D Surfaces)

Programming procedures using wire frame, splines, and surface modeling. Rough, finish, and high speed machining. Editing, post-processing and verifying programs.

MCNC 76M: CAD/CAM Mastercam (Advanced)

Mastercam; complex surfacing for milling machines and contouring surfaces for lathes. Tooling, workflow and programming for horizontal machining centers

De Anza College
21250 Stevens Creek Boulevard
Cupertino, CA 95014
www.deanza.edu

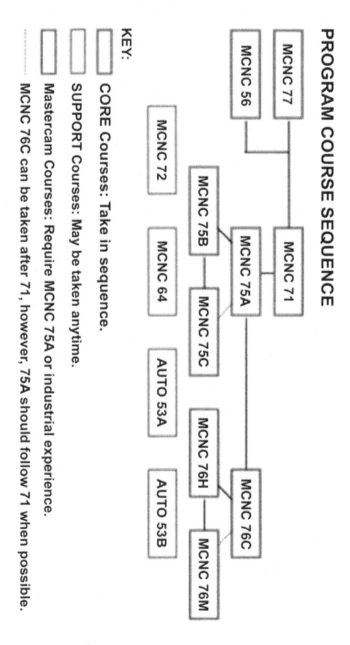

PROGRAM COURSE SEQUENCE

KEY:
- CORE Courses: Take in sequence.
- SUPPORT Courses: May be taken anytime.
- Mastercam Courses: Require MCNC 75A or industrial experience.

MCNC 76C can be taken after 71, however, 75A should follow 71 when possible.

Made in the USA
Coppell, TX
26 August 2022